AF683281

Numerical Solution of Ordinary Differential Equations

Numerical Solution of Ordinary Differential Equations

Editor

Nik Pachis

scitus
academics

Numerical Solution of Ordinary Differential Equations
Edited by **Nik Pachis**

ISBN: 978-1-68117-448-8
Library of Congress Control Number: 2016937608

© 2017 by
SCITUS Academics LLC,
www.scitusacademics.com
Box No. 4766, 616 Corporate Way,
Suite 2, Valley Cottage,
NY 10989

This book contains information obtained from highly regarded resources. Copyright for individual articles remains with the authors as indicated. All chapters are distributed under the terms of the Creative Commons Attribution License, which permits unrestricted use, distribution, and reproduction in any medium, provided the original author and source are credited.

Notice

Reasonable efforts have been made to publish reliable data and views articulated in the chapters are those of the individual contributors, and not necessarily those of the editors or publishers. Editors or publishers are not responsible for the accuracy of the information in the published chapters or consequences of their use. The publisher believes no responsibility for any damage or grievance to the persons or property arising out of the use of any materials, instructions, methods or thoughts in the book. The editors and the publisher have attempted to trace the copyright holders of all material reproduced in this publication and apologize to copyright holders if permission has not been obtained. If any copyright holder has not been acknowledged, please write to us so we may rectify.

Printed in United States of America on Acid Free Paper

Preface

Numerical methods are used to find numerical approximations to common differential equations (ODE) solutions for ordinary differential equations. An ordinary differential equation or ODE is a differential equation containing one or more independent variable functions and derivatives. In contrast to the term partial differential equation, the term "ordinary" can be used for more than one independent variable. Ordinary differential equations in science and engineering are ubiquitous: from the first examples (Newton, Leibniz, Euler, and Lagrange), in geometry and mechanics, in chemical reaction kinetics, molecular dynamics, electronic circuits, population dynamics and much more. They also lead to numerical treatment of time-dependent partial differential equations, which after semi-discretization in space are even more impressively ubiquitous in our technologically developed and financially controlled world.

This book provides a comprehensive and easy introduction to classical themes in the numerical solution of ordinary differential equations. The book approach not only explains the presented mathematics, but also helps readers understand how these numerical methods are used in the real-world to solve problems.

Table of Contents

CHAPTER 1

Tau-Collocation Approximation Approach for Solving First and Second Order Ordinary Differential Equations

James E. Mamadu[1], Ignatius N. Njoseh[2]

[1]Department of Mathematics, University of Ilorin, Ilorin, Nigeria

[2]Department of Mathematics and Computer Science, Delta State University, Abraka, Nigeria

ABSTRACT

This paper presents Tau-collocation approximation approach for solving first and second orders ordinary differential equations. We use the method in the stimulation of numerical techniques for the approximate solution of linear initial value problems (IVP) in first and second order ordinary differential equations. The resulting numerical evidences show the method is adequate and effective.

KEYWORDS

Ordinary Differential Equation (ODE), Initial Value Problem (IVP), Canonical Polynomial, Collocation

1. INTRODUCTION

The subject of Ordinary Differential Equation (ODE) is an important aspect of mathematics. It is useful in modeling a wide variety of physical phenomena—chemical reactions, satellite orbit, electrical networks, and so on. In many cases, the independent variable represents time so that the differential equation describes changes, with respect to time, in the system being modeled. The solution of the equation will be a representation of the state of the system. Consequently, the problem of finding the solution of a differential equation plays a significant role in scientific research, particularly, in the stimulation of physical phenomena. However, it is usually impossible to obtain direct solution of differential equations for systems to be modeled, especially complex ones encountered in real world problems. Since most of these equations are, or can be approximated by ordinary differential equations, a fast, accurate and efficient ODE solver is much needed. The Tau method was introduced by [1] to provide approximate polynomial solution for linear ordinary differential equation with polynomial coefficient.

The method takes advantage of the special properties of Chebychev polynomials. The main idea is to obtain an approximate solution of a given problem by solving an approximate problem. To further enhance the desired simplicity Lanczos introduced the systematic use of the canonical polynomials in the Tau method. The difficulties presented by the construction of such polynomials limited its application to first order ODE with the polynomial coefficient. The said difficulties were resolved by [2] when he proposed the generation of these canonical polynomials recursively. The beauty of the result of Ortiz is that the elements of canonical polynomials sequences by means of a simple recursive relation which is self starting and explicit. There are many literature developed concerning the Tau/Tau-collocation approximation method (see ([3] - [7])).

In this paper, we apply the Tau-collocation approximation method for the solution of linear initial value problems of the first and second order ODE in its differential and canonical form. We perform some numerical stimulation on some selected problems and compare the performance/effectiveness of the method with the analytic solutions given.

2. THE TAU METHOD

Lanczos [1] approximated a solution $y(x)$ of the differential

$$\sum_{r=o}^{m} P_r(x) y^{(r)} = f(x), \ 0 \le x \le 1 \tag{1}$$

where $P_r(x)$, $f(x)$ are polynomials. $y^{(r)}(x)$ denotes the rth order derivative of $y(x)$ with respect to x and $y^{(0)}(x)$ taken simply as $y(x)$ by a polynomial

$$y_n(x) = \sum_{r=o}^{n} a_r x^r, \ n < \infty \tag{2}$$

and determines the coefficient $a_r, r = 0(1)n$ of (2) such that $y_n(x)$ satisfies (1) perturbed by a term(s), which are calculated as part of the process. That is, $y_n(x)$ satisfies

$$\sum_{r=0}^{m} P_r(x) y^{(r)} = f(x) + H_n(x). \tag{3}$$

Then

$$H_n(x) = \sum_{r=0}^{m+s-1} {}_{m+s-r} T^{*}_{n-m+r+1}, \quad 0 \leq x \leq 1 \tag{4}$$

where m is the order of the differential equation, s is the number of over-determination, $r = 1(1)m+s$, are the parameters to be determined, and

$$T_r^{*}(x) = \mathrm{Cos}\left[r\mathrm{Cos}^{-1}\left(\frac{2x-a-b}{b-a}\right)\right] \equiv \sum_{k=0}^{r} C_k^{(r)} x^k \tag{5}$$

is the rth degree shifted Chebychev polynomial valid in the interval $[a,b]$ (assuming (1) is defined in this interval).

The free parameters in Equation (4) and the coefficient a_r, $r \in 0(1)n$ in (2) are obtained by equating the values of x in (3) together with (1) to zero.

2.1. Description of the Differential Form

Considering the mth order linear differential Equation ([1] [2])

$$L(y(x)) := \sum_{r=0}^{m} P_r(x) y^{(r)} = f(x), \tag{6}$$

$$y(0) = \alpha_0, y'(0) = \alpha_1, \cdots, y^{(m-1)}(0) = \alpha_{(m-1)},$$

with y(x) as the exact solution in $a \le x \le b, |a| < \infty, |b| < \infty.$

We seek an approximate solution of the differential solution by the Tau method using the nth degree polynomial function

$$y_n(x) = \sum_{r=0}^{m} P_r(x) x^r, n < \infty \tag{7}$$

which satisfies the perturbed problem

$$L(y(x)) := \sum_{r=0}^{m} P_r(x) y^{(r)} = f(x) + H_n(x) \tag{8}$$

We equate the corresponding coefficient of x in (8) and using the initial conditions

$$y(0) = \alpha_0, y'(0) = \alpha_1, \cdots, y^{(m-1)}(0) = \alpha_{(m-1)},$$

We then solve the system of equation by Gaussian elimination method.

2.2. Collocation Approach to the Tau Method

The Lanczos Tau method in [4] is an economization process for a function that is implicitly defined by differential equation. Let us assume an approximation of the power series expansion $u(x)$ as

$$u_n(x)=\sum_{r=0}^{n} a_r(x)x^r \tag{9}$$

Consider an approximation to the residual $R_n(x)$ as

$$R_n(x)={}_1T_n^*(x)+{}_2T_{n-1}^*(x)+\cdots+{}_mT_{n-m+1}^*(x) \tag{10}$$

Then by the Tau method, if

$$L(u(x))=f(x) \tag{11}$$

we have

$$L(u(x))=f(x)+R_n(x)=f(x)+{}_1T_n^*(x)+{}_2T_{n-1}^*(x)+\cdots+{}_mT_{n-m+1}^*(x) \tag{12}$$

where L is a linear differential operator of order n.

We collocate (12) at $x_k=kh, k=1(1)N,$ where $h=\dfrac{b-a}{n+1}$ to have

$$L(u(x_k))=f(x_k)+R_n(x_k)=f(x_k)+{}_1T_n^*(x_k)+{}_2T_{n-1}^*(x_k)+\cdots+{}_mT_{n-m+1}^*(x_k) \tag{13}$$

The parameter $1,2,\cdots,m$ may be eliminated leaving the unknown coefficient $a_k, k=0(1)n$ with $(n+1)$ linear equations which can be solved by Gaussian elimination.

3. ERROR ESTIMATION

Let us in this section consider and obtain the error estimator for the approximate solution of (1) and (9). Let $E_n = u(x) - u_n(x)$ be the error function of $u_n(x)$ to $u(x)$, where $u(x)$ is the exact solution of (1) and (9). Therefore $u_n(x)$ satisfies the perturbed problems:

$$L(u_n(x)) := \sum_{r=0}^{m} P_r(x)(u_n(x))^{(r)} = f(x) + H_n(x), x \in [a,b], |a| < \infty, |b| < \infty, n < \infty \tag{14}$$

And

$$L(u_n(x)) := \sum_{r=0}^{m} a_r x^r = f(x) + H_n(x) \tag{15}$$

where H_n is uniquely defined as in (4).

To obtain the perturbation term $H_n(x)$, we substitute the computed solution $u_n(x)$ such that

$$L(u_n(x)) := \sum_{r=0}^{m} P_r(x)(u_n(x))^{(r)} - f(x) = H_n(x)$$

and

$$L(u_n(x)) := \sum_{r=0}^{m} a_r x^r - f(x) = H_n(x)$$

We then proceed to find an approximate $E_{n,N}(x)$ to the error function $E_n(x)$ in the same manner as we did for the solution of (1) and (9).

Thus, the error function, E_n, satisfy the problem

$$L(E_n) := \sum_{r=0}^{m} P_r(x) E_n^r - f(x) = -H_n(x) \tag{16}$$

And

$$L(E_n) := \sum_{r=0}^{m} a_r x^r - f(x) = 0 \tag{17}$$

which satisfies the conditions prescribed.

4. ILLUSTRATIVE EXAMPLES

In this section, two initial value problems are considered to show the efficiency of the method.

Example 1

Consider linear initial value problem in second order ordinary differential equation

$$\begin{cases} u''(x)+u(x)=0 \\ u(0)=1, u'(1)=3 \end{cases} \tag{18}$$

We solve [4] for $n=2$ using; (i) The Tau method; and (ii) Tau-collocation method.

The analytic solution is $u(x)=\text{Cos}x-\dfrac{\text{Sin}x(\text{Cos}1-3)}{\text{Sin}1}$.

By the Tau method we obtain the linear differential operator as

$$L\equiv\left(\frac{d^2}{dx^2}+1\right) \tag{19}$$

The associated canonical polynomials are obtained as follows:

$$Lx^r=\left(\frac{d^2}{dx^2}+1\right)x^r \tag{20}$$

$$Lx^r=r(r-1)LQ_{r-2}(x)+LQ_r(x)$$

$$L^{-1}\cdot Lx^r=L^{-1}\cdot L\left[r(r-1)Q_{r-2}(x)+Q_r(x)\right]$$

$$\Rightarrow x^r=r(r-1)Q_{r-2}(x)+Q_r(x)$$

$$\Rightarrow Q_r(x)=x^r-r(r-1)Q_{r-2}(x)$$

The canonical polynomials, $Q_r(x), r \geq 0$, obtained here can easily be obtained from [3] where the generalized form of the canonical polynomials was reported.

For $r = 0(1)2$, we have $Q_0(x) = 1, Q_1(x) = x$ and $Q_2(x) = x^2 - 2$

These polynomials are substituted into Equation (12) to give

$$Lu_n(x) = {}_1T_n^*(x) + {}_2T_{n-1}^*(x) \tag{21}$$

Using Equation (5),

$$Lu_n(x) = {}_1\sum_{r=0}^{n} C_r^{(n)} x^r + {}_2\sum_{r=0}^{n-1} C_r^{(n-1)} x^r(x)$$

$$Lu_n(x) = {}_1\sum_{r=0}^{n} C_r^{(n)} LQ_r(x) + {}_2\sum_{r=0}^{n-1} C_r^{(n-1)} LQ_r(x)$$

$$u_n(x) = {}_1\sum_{r=0}^{n} C_r^{(n)} Q_r(x) + {}_2\sum_{r=0}^{n-1} C_r^{(n-1)} Q_r(x) \tag{22}$$

Since

$$u_2(x) = {}_1\sum_{r=0}^{2} C_r^{(2)} Q_r(x) + {}_2\sum_{r=0}^{1} C_r^{(n-1)} Q_r(x)$$

Now,

$T_0(x) = 1, T_1(x) = 2x - 1$ and $T_2(x) = 8x^2 - 8x + 1$

$$u_2(x) = {}_1\left(C_0^{(2)} Q_0(x) + C_1^{(2)} Q_1(x) + C_2^{(2)} Q_2(x)\right) + {}_2\left(C_0^{(1)} Q_0(x) + C_1^{(1)} Q_1(x)\right)$$

$$\Rightarrow u_2\left(x\right) = {}_1\left(1-8x+8\left(x^2-2\right)\right) + {}_2\left(-1+2x\right) \tag{23}$$

Using initial conditions on Equation (23) and simplifying further we get the approximate solution as

$$u_2\left(x\right) = 1 + \frac{46}{15}x - \frac{16}{15}x^2 .$$

Considering the Tau-collocation method we have:

Let

$$u_2\left(x\right) = \sum_{r=0}^{2} a_r\left(x\right)x^r \tag{24}$$

$$u_2\left(x\right) = a_0 + a_1 x + a_2 x^2$$

$$u_2'\left(x\right) = a_1 + 2a_2 x$$

$$u_2''\left(x\right) = 2a_2$$

Substituting into (13) we have,

$$2a_2 + a_0 + a_1 x + a_2 x^2 = {}_1T_2^*\left(x\right) + {}_2T_1^*\left(x\right) \tag{25}$$

$$2a_2 + a_0 + a_1 x + a_2 x^2 = {}_1\left(1-8x+8\left(x^2-2\right)\right) + {}_2\left(-1+2x\right)$$

Now collocating at $x_k = kh, h = \dfrac{b-a}{n+1}, k = 1,2,3$ and using the initial conditions, we obtain the approximate solution as

$$u_2(x) = 1 + \frac{46}{15}x - \frac{16}{15}x^2$$

Example 2

Consider the first order IVP

$$Ly(x) = (1+x)y'(x) + y(x) = 0,\ y(0) = 1 \tag{26}$$

The exact solution is $y(x) = \frac{1}{x+1}$

For the given IVP, we can deduce that $m = 1$ and $s = 0.$

The differential formulation is as follows:

Let

$$y_n(x) = \sum_{r=o}^{n} a_r x^r \tag{27}$$

Taking $n = 5$

$$Ly_5(x) = \sum_{r=o}^{m+s-1} {}_{m+s-r}T^*_{n-m+r+1}(x)$$

$$\Rightarrow Ly_5(x) = \sum_{r=o}^{0} {}_1T^*_5$$

Where

$$T_5^* = \sum_{r=o}^{5} C_r^{(5)} x^r \tag{28}$$

Hence

$$(1+x)\,y_5'(x) + y_5(x) = {}_1T_5^*(x) \tag{29}$$

But

$$Ly_5(x) = {}_1\sum_{r=o}^{5} C_r^{(5)} x^r$$

Using (28) and (30) in (29) we obtain,

$$(x+1)\sum_{r=o}^{5} r a_r x^{r-1} + \sum_{r=o}^{5} r a_r x^r = {}_1\sum_{r=o}^{5} C_r^{(5)} x^r \tag{31}$$

Expanding and equating coefficients of powers of x, the resulting linear equations together with the equations obtained using the initial conditions is written in the form,

$$A\underline{x} = \underline{b}$$

where

$$\begin{pmatrix} 1 & 1 & 0 & 0 & 0 & 0 & \cdots & C_0^{(5)} \\ 0 & 2 & 2 & 0 & 0 & 0 & \cdots & C_1^{(5)} \\ 0 & 0 & 3 & 3 & 0 & 0 & \cdots & C_2^{(5)} \\ 0 & 0 & 0 & 4 & 4 & 0 & \cdots & C_3^{(5)} \\ 0 & 0 & 0 & 0 & 5 & 5 & \cdots & C_4^{(5)} \\ 0 & 0 & 0 & 0 & 0 & 0 & \cdots & C_5^{(5)} \\ 1 & 0 & 0 & 0 & 0 & 0 & \cdots & 0 \end{pmatrix}$$

$$\underline{x} = \left(a_0, a_1, a_2, a_3, a_4, a_5, 1\right)^{\mathrm{T}},$$

$$\underline{b} = \left(0,0,0,0,0,0,1\right)^{\mathrm{T}}$$

Using Equation (5), we obtain the following values,

$$C_0^{(5)} = -1, C_1^{(5)} = 50, C_2^{(5)} = -400, C_3^{(5)} = 1120, C_4^{(5)} = -1280 \text{ and } C_5^{(5)} = 512.$$

Using these values in the matrix and solving by Gaussian elimination method, we have,

$$a_0 = 1, a_1 = -\frac{2339}{2342}, a_2 = \frac{1132}{1171}, a_3 = -\frac{932}{1171}, a_4 = \frac{512}{1171}, a_5 = -\frac{128}{1171}, {}_1 = \frac{3}{2342}$$

The approximate solution is:

$$y_5(x) = \frac{1}{2342}\left(2342 - 2339x + 2264x^2 - 1864x^3 + 1024x^4 - 256x^5\right)$$

5. DISCUSSION OF RESULTS

The results obtained above show that the Tau-collocation method is appropriate for the solution of linear initial value problems of first and second kind ordinary differential equations. From the tables (Table 1 and Table 2) of results presented above, we observe that the approximate solution considered at grid points, $n = 2$ and $n = 5$, for examples 1 and 2 converges to the analytic solution with maximum absolute errors of 10^{-3} and 10^{-5} respectively. We obtain satisfactory results because of the excellent convergence rate of the Tau-collocation approximation method, which is very close to the minimax polynomial which minimizes the maximum error in approximation. Thus, the approximate solution will match the analytic solution as n increases.

Table 1. Numerical results for example 1.

X	*Exact Solution*	*Approximate solution,* $n = 2$	*Error*
0.1	1.2868265	1.2960000	9.1735e−03
0.2	1.5607954	1.5706667	9.8712e−03
0.3	1.8191694	1.8240000	4.8306e−03
0.4	2.0593669	2.0560000	3.3669e−03
0.5	2.2789879	2.2666667	1.2321e−02
0.6	2.4758379	2.4560000	1.9838e−02
0.7	2.6479502	2.6240000	2.3950e−02
0.8	2.7936051	2.7706667	2.2938e−02
0.9	2.9113471	2.8960000	1.5347e−02
1.0	3.0000000	3.0000000	0.0000e+00

Table 2. Numerical results for Example 2.

X	*Exact Solution*	*Approximate solution, $n = 2$*	*Error*
0.1	0.9090909	0.9090418	4.19133e–05
0.2	0.8333333	0.8332214	1.1195e–04
0.3	0.7692308	0.7691735	5.7276e–05
0.4	0.7142857	0.7143198	3.4081e–05
0.5	0.6666667	0.6667378	7.1164e–05
0.6	0.6250000	0.6250298	2.9821e–05
0.7	0.5882353	0.5881915	4.3800e–05
0.8	0.5555556	0.5554809	7.4633e–05
0.9	1.5263158	0.5265873	2.8445e–05
0.10	0.5000000	0.5000000	0.0000e+00

6. CONCLUSIONS

This paper has considered Tau-collocation approximation approach for solving particular first and second order ordinary differential equations. The method offers several advantages which include, among others:

1) It takes advantages of the special properties of Chebychev polynomials which can be easily generated recursively;

2) Elements of canonical polynomials sequences by means of a simple re-cursive relation which is self starting and explicit; and

3) It can easily be programmed for experimentation.

Tau-Collocation method can be extended to higher order ordinary differential equations and stochastic differential equations. It can also be used to solve integro-differential and stochastic integro-differential equations.

REFERENCES

1. Lanczos, C. (1956) Applied Analysis. Prentice-Hall, Engle-Wood Cliffs, New Jersey.

2. Ortiz, E.L. (1969) The Tau Method. SIAM Journal on Numerical Analysis, 6, 480-492.

3. Yisa, B.M. and Adeniyi, R.B. (2012) A Generalized Formulation for Canonical Polynomials for M-Th Order Non Over-Determined Ordinary Differential Equations. Internal Journal of Engineering Research and Technology. www.ijert.org

4. Coleman, J.P. (1974) Lanczos Tau Method. IMA Journal of Applied Mathematics, 7, 85-97.

5. El-Daou, M.K., Ortiz, E.L. and Samara, H. (1993) A Unified Approach to the Tau Method and Chebychev Series Expansion Technique. Computer and Mathematics with Applications, 25, 73-82.

6. Ortiz, E.L. (1975) Step by Step Tau-Method—Part 1. Computer and Mathematics with Applications, 1, 381-392.

7. Khajah, H. (1999) Tau-Method Approximation of a Generalized Epstein-Hubbel Elliptic Type Integral. Computer and Mathematics, 68, 1615-1621.

CHAPTER 2

Variational Iteration Method Solutions for Certain Thirteenth Order Ordinary Differential Equations

Tunde A. Adeosun[1], Olugbenga J. Fenuga[2], Samuel O. Adelana[3], Abosede M. John[1], Ogunjimi Olalekan[1], Kazeem B. Alao[1]

[1]Department of Mathematics, Yaba College of Technology, Lagos, Nigeria

[2]Department of Mathematics, University of Lagos, Lagos, Nigeria

[3]Department of Mechanical Engineering, Yaba College of Technology, Lagos, Nigeria

ABSTRACT

In this paper, we extend variational iteration method (VIM) to find approximate solutions of linear and nonlinear thirteenth order differential equations in boundary value problems. The method is based on boundary valued problems. Two numerical examples are presented for the numerical illustration of the method and their results are compared with those considered by [1,2]. The results reveal that VIM is very effective and highly promising in comparison with other numerical methods.

KEYWORDS

Variational Iteration Method; Boundary Value Problems; Linear and Nonlinear Problems; Approximate Solution

1. INTRODUCTION

Boundary Value Problems (BVPs) in linear and nonlinear differential equations has been one of the major problems. Many phenomena in applied Mathematics and other sciences can be described very successfully by models using mathematical tools from ordinary differential equations. [3-5] claimed that ODE has gained much attention recently due to exact description of nonlinear phenomena. In the past, both Mathematicians and Physicists have made significant progress in this direction. Analytical methods that are commonly used to solve nonlinear equations are very restricted and numerical methods involving description of variables on the other hand give rise to rounding off errors. Since these equations are linear and nonlinear, it may not have precise analytical solutions. On the other hand, solving these differential equations analytically may guide Mathematicians to know how to describe some physical process deeply and sometimes lead to know some facts which are not simply understood through common observations. As a result, these equations have to be solved using other methods.

References [6,7] gave examples on the free vibration analysis of beam structure that is governed by a fourthorder differential equation and ring structured by a sixthorder differential equation. Moreover, when considering the instability setting in an infinite horizontal layer of fluid, which is heated from below and is subject to the action of rotation, we may model the instability as ordinary convention and over stability by a sixth-order Ordinary Differential Equation (ODE) and an eighth-order ODE respectively. Even higher order ODE can be involved when a uniform magnetic field is applied across the fluid in the same direction as gravity. Ordinary convention and over stability yield a tenth order and a twelfth order ODEs respectively. Such problems modeled by high-order differential equations are often associated with initial boundary conditions, so that the problem is well posed. References [8,9] provided examples on the beam vibration problem, which is governed by a fourth-order differential equation, has four boundary conditions. It is noted that for well-posed problems, the number of boundary conditions is the same as the order of the differential equation and in general they are only given at each boundary [10-12].

In this present work, Variational Iteration Techniques for the solution of thirteen order boundary value problem is studied, following the variational iteration techniques for the solution of eight order boundary value problems.

2. VARIATIONAL ITERATION ANALYSIS

To illustrate the basic concept of the variational iteration method, we consider the following general system.

$$Lu + Nu = f(x) \tag{1}$$

where L is a linear operator, N a nonlinear operator and $f(x)$ is a forcing term. Following the variational iteration technique used in [13-17], a correct functional can be constructed as

$$u_{n+1}(x)=u_n(x)+\int_{x_0}^{x}\lambda\left(Lu_n(s)+Nu_n(s)-f(s)\right)\mathrm{d}x \tag{2}$$

where λ is a Lagrange multiplier, which can be identified optimally via variational iteration method. It may be mentioned that n denotes the nth approximation u_n is considered to be restricted variation i.e. $\delta u_n=0$. It may also be noted that the Equation (2) is known as a correct functional.

Due to the exert identification of the Lagrange multiplier, the solution of the linear problems can be determined in a single iteration step. The following system of differential equations helps in understanding the variational iteration method.

$$x_i(t)=g_i(t_i,x_i),\ \ i=1,2,\cdots,n \tag{3}$$

along in the boundary conditions $x_i(0)=c_i, i=1,2,\cdots,n$.

The system (3) is rewritten as

$$x_i(t)=g_i(x_i)+f_i(t),\ \ i=1,2,\cdots,n$$

The following system is used to approximate the correct functional for the nonlinear system. (see (4) below)

where $\lambda_1=\pm 1,\ \ i=1,2,\cdots,n$ are Lagrange multipliers, while $x_1, x_2, \cdots, x_n$ denote the restricted variations.

The following iterative scheme can be obtained for $\lambda_i=-1, i=1,2,\cdots,n$ as (see (5) below)

Considering the initial guess values to be $x_i(0) = c_i, i = 1, 2, \cdots, n,$

The final approximation solution can be written as $x_i(t) = \lim_{k \to \infty} x_i^{(k)}(t)$ can be terminated after suitable number of iterations.

3. MATHEMATICAL FORMULATION

The basic motivation of this research work is to apply the variational iteration technique to solve a system of differential equations. The boundary value problem, to be solved is as follows

$$\left.\begin{array}{ll} y^{(13)}(x) + f(x)y(x) = g(x), & x \leftarrow [a,b] \\ y(a) = A_1, & y(b) = B_1, \\ y^{(1)}(a) = A_2, & y^{(1)}(b) = B_2, \\ y^{(2)}(a) = A_3, & y^{(2)}(b) = B_3, \\ y^{(3)}(a) = A_4, & y^{(3)}(b) = B_4, \\ y^{(4)}(a) = A_5, & y^{(4)}(b) = B_5, \\ y^{(5)}(a) = A_6, & y^{(5)}(b) = B_6, \\ y^{(0)}(a) = C_1, & \end{array}\right\} \quad (6)$$

where $A_i, B_i, i = 1(1)6$ and c_1 are the finite real constants and functions f is continuous on $[a,b]$.

4. NUMERICAL EXAMPLES

To implement the method developed, two examples are considered.

Example 1

For $x \in [0,1]$, the following linear boundary value problem is considered

$$x_1^{(k+1)}(t) = x_1^{(0)}(t) + \int_{x_0}^{t} \lambda_1 \left(x_1^{(k)}(T), g_1\left(x_1^{(k)}(T), x_2^{(k)}(T), \cdots, x_n^{(k)}(T) \right) - f_1(T) \right) dT$$

$$x_2^{(k+1)}(t) = x_2^{(0)}(t) + \int_{x_0}^{t} \lambda_2 \left(x_2^{(k)}(T), g_2\left(x_1^{(k)}(T), x_2^{(k)}(T), \cdots, x_n^{(k)}(T) \right) - f_2(T) \right) dT$$

$$\vdots$$

$$x_n^{(k+1)}(t) = x_n^{(0)}(t) \;\; t\Big|_{x_0}^{t} \lambda_1 \left(x_n^{(k)}(T), g_n\left(x_1^{(k)}(T), x_2^{(k)}(T), \cdots, x_n^{(k)}(T) \right) - f_n(T) \right) dT$$

(4)

$$x_1^{(k+1)}(t) = x_1^{(0)}(t) - \int_{x_0}^{t} \left(x_1^{(k)}(T), g_1\left(x_1^{(k)}(T), x_2^{(k)}(T), \cdots, x_n^{(k)}(T) \right) - f_1(T) \right) dT$$

$$x_2^{(k+1)}(t) = x_2^{(0)}(t) - \int_{x_0}^{t} \left(x_2^{(k)}(T), g_2\left(x_1^{(k)}(T), x_2^{(k)}(T), \cdots, x_n^{(k)}(T) \right) - f_2(T) \right) dT$$

$$\vdots$$

$$x_n^{(k+1)}(t) = x_n^{(0)}(t) - \int_{x_0}^{t} \left(x_n^{(k)}(T), g_n\left(x_1^{(k)}(T), x_2^{(k)}(T), \cdots, x_n^{(k)}(T) \right) - f_n(T) \right) dT$$

(5)

$$\left.\begin{aligned}
&y^{(13)}(x)=\cos x-\sin x,\\
&y^{(0)}(0)=1,\\
&y^{(1)}(0)=1,\\
&y^{(2)}(0)=-1,\\
&y^{(3)}(0)=-1,\\
&y^{(4)}(0)=1,\\
&y^{(5)}(0)=1,\\
&y^{(6)}(0)=-1,\\
&y^{(0)}(1)=1,\\
&y^{(1)}(1)=-1,\\
&y^{(2)}(1)=-1,\\
&y^{(3)}(1)=1,\\
&y^{(4)}(1)=1,\\
&y^{(5)}(1)=-1,
\end{aligned}\right\} \tag{7}$$

The analytical solution of the given problem from [8]

$$y(x)=\sin x+\cos x \tag{8}$$

The given thirteen order BVP can be transferred with the following system

$$\left.\begin{aligned}
\frac{dy}{dx} &= p(x) \\
\frac{dp}{dx} &= q(x) \\
\frac{dq}{dx} &= r(x) \\
\frac{dr}{dx} &= s(x) \\
\frac{ds}{dx} &= m(x) \\
\frac{dm}{dx} &= n(x) \\
\frac{dn}{dx} &= l(x) \\
\frac{dl}{dx} &= j(x) \\
\frac{dj}{dx} &= u(x) \\
\frac{du}{dx} &= w(x) \\
\frac{dw}{dx} &= g(x) \\
\frac{dg}{dx} &= z(x) \\
\frac{dz}{dx} &= \cos x - \sin x
\end{aligned}\right\} \qquad (9)$$

$y(0)=1,\ p(0)=1,\ q(0)=-1,\ r(0)=-1,\ s(0)=+1,$

$m(0)=1,\ n(0)=1, l(0)=a,\ j(0)=b,$

$u(0)=c,\ w(0)=d,\ g(0)=e,\ z(0)=f$

This system of differential equation can be written in terms of the following system of integral equation with Lagrange multipliers $\lambda_i = +1,\ i = 1,2,3,\cdots,\beta$

$$\left.\begin{aligned}
y^{k+1}(x) &= 1 + \int_0^x p^{(k)}(t)\,\mathrm{d}t \\
p^{k+1}(x) &= 1 + \int_0^x q^{(k)}(t)\,\mathrm{d}t \\
q^{k+1}(x) &= -1 + \int_0^x r^{k}(t)\,\mathrm{d}t \\
r^{k+1}(x) &= -1 + \int_0^x s^{k}(t)\,\mathrm{d}t \\
s^{k+1}(x) &= 1 + \int_0^x m^{k}(t)\,\mathrm{d}t \\
m^{k+1}(x) &= 1 + \int_0^x n^{k}(t)\,\mathrm{d}t \\
n^{k+1}(x) &= -1 + \int_0^x l^{k}(t)\,\mathrm{d}t \\
l^{k+1}(x) &= a + \int_0^x j^{k}(t)\,\mathrm{d}t \\
j^{k+1}(x) &= b + \int_0^x u^{k}(t)\,\mathrm{d}t \\
u^{k+1}(x) &= c + \int_0^x w^{k}(t)\,\mathrm{d}t \\
w^{k+1}(x) &= d + \int_0^x g^{k}(t)\,\mathrm{d}dt \\
g^{k+1}(x) &= e + \int_0^x z^{k}(t)\,\mathrm{d}t \\
z^{k+1}(x) &= f + \int_0^x y^{k}(t)\,\mathrm{d}t \\
z^{k+1}(x) &= f + \int_0^x (\cos t - \sin t)\,\mathrm{d}t
\end{aligned}\right\} \qquad (10)$$

Where

$$y^{(0)}(x)=1,\ p^{(0)}(x)=1,\ q^{(0)}(x)=-1,\ r^{(0)}(x)=-1,$$
$$s^{(0)}(x)=1,\ m^{(0)}(x)=1,\ n^{(0)}(x)=-1,$$
$$l^{(0)}(x)=a,\ j^{(0)}(x)=b,\ u^{(0)}(x)=c,\ w^{(0)}(x)=d,$$
$$g^{(0)}(x)=e,\ z^{(0)}(x)=f$$

The system of Equation (11) will provides the following approximations

When $k=0$

$$y^{(1)}(x)=1+\int_0^x \mathrm{d}x=1+x$$

$$p^{(1)}(x)=1+\int_0^x(-\mathrm{d}x)=1-x$$

$$q^{(1)}(x)=-1+\int_0^x(-\mathrm{d}x)=-1-x$$

$$r^{(1)}(x)=-1+\int_0^x \mathrm{d}x=-1+x=x-1$$

$$s^{(1)}(x)=1+\int_0^x \mathrm{d}x=1+x$$

$$m^{(1)}(x)=1+\int_0^x -\mathrm{d}x=1-x$$

$$n^{(1)}(x)=-1+\int_0^x a\mathrm{d}x=-1+ax=ax-1$$

$$l^{(1)}(x)=a+\int_0^x b\mathrm{d}x=a+bx$$

$$j^{(1)}(x) = b + \int_0^x c\mathrm{d}x = b + cx$$

$$u^{(1)}(x) = c + \int_0^x d\mathrm{d}x = c + dx$$

$$w^{(1)}(x) = d + \int_0^x e\mathrm{d}x = d + ex$$

$$g^{(1)}(x) = e + \int_0^x f\mathrm{d}x = e + fx$$

$$z^{(1)}(x) = f + \sin x + \cos x$$

$$k = 1$$

$$y^{(2)}(x) = 1 + \int_0^x p^1\mathrm{d}x = 1 + \int_0^x (1 - x)\mathrm{d}x = 1 + x - \frac{x^2}{2}$$

$$p^{(2)}(x) = 1 + \int_0^x q^1\mathrm{d}x = 1 + \int_0^x (-1 - x)\mathrm{d}x = 1 - x - \frac{x^2}{2}$$

$$q^{(2)}(x) = -1 + \int_0^x r^1\mathrm{d}x = -1 + \int_0^x (x - 1)\mathrm{d}x = -1 + \frac{x^2}{2} - x$$

$$= \frac{x^2}{2} - x - 1$$

$$r^{(2)}(x) = -1 + \int_0^x s^1\mathrm{d}x = -1 + \int_0^x (1 + x)\mathrm{d}x = -1 + x + \frac{x^2}{2}$$

$$= \frac{x^2}{2} + x - 1$$

$$s^{(2)}(x)=1+\int_0^x m^1 \mathrm{dx}=1+\int_0^x (1-x)\mathrm{dx}=1+x-\frac{x^2}{2}$$

$$m^{(2)}(x)=1+\int_0^x n^1 \mathrm{dx}=1+\int_0^x (ax-1)\mathrm{dx}=1+\frac{ax^2}{2}-x$$

$$n^{(2)}(x)=-1+\int_0^x L^1 \mathrm{dx}=-1+\int_0^x (a+bx)\mathrm{dx}$$

$$=-1+ax+\frac{bx^2}{2}$$

$$L^{(2)}(x)=a+\int_0^x j^1 \mathrm{dx}=a+\int_0^x (b+cx)\mathrm{dx}=a+bx+\frac{cx^2}{2}$$

$$j^2(x)=b+\int_0^x u^1 \mathrm{dx}=b+\int_0^x (c+dx)\mathrm{dx}=b+cx+\frac{dx^2}{2}$$

$$u^2(x)=c+\int_0^x w^1 \mathrm{dx}=c+\int_0^x (d+ex)\mathrm{dx}=c+dx+\frac{ex^2}{2}$$

$$w^2(x)=d+\int_0^x g^1 \mathrm{dx}=d+\int_0^x (e+fx)\mathrm{dx}=d+ex+\frac{fx^2}{2}$$

$$g^{(2)}(x)=e+\int_0^x z^1 \mathrm{dx}=e+\int_0^x (f+\sin x+\cos x)\mathrm{dx}$$

$$=e+fx-\cos x+\sin x=e+fx+\sin x-\cos x$$

$$z^{(2)}(x)=f+\int_0^x (\cos x-\sin x)\mathrm{dx}=f+\sin x+\cos x$$

$$y^{(3)}(x)=1+\int_0^x p^2\mathrm{d}x=1+\int_0^x\left(1-x-\frac{x^2}{2}\right)\mathrm{d}x$$
$$=1+x-\frac{x^2}{2}+\frac{x^3}{6}.$$

$$y^{11}(x)=1+x-\frac{x^2}{2}-\frac{x^3}{6}+\frac{x^4}{24}-\frac{x^5}{120}-\frac{x^6}{720}+\frac{ax^7}{5040}$$
$$+\frac{bx^8}{40320}+\frac{cx^9}{362880}+\frac{dx^{10}}{3628800}+\frac{ex^{11}}{39916800}$$

The solution is given as

$$y(x)=1+x-\frac{x^2}{2}-\frac{x^3}{6}+\frac{x^4}{24}-\frac{x^5}{120}-\frac{x^6}{720}+\frac{ax^7}{5040}+\frac{bx^8}{40320}$$
$$+\frac{cx^9}{362880}+\frac{dx^{10}}{3628800}+\frac{ex^{11}}{39916800} \tag{11}$$

The coefficients a, b, c, d and e are obtained using MAPLE 13 with boundary condition at $x=1$, which gives

$$a=1, b=1, c=-1, d=0.999997, e=-1$$

The series solution can thus, be written as

$$y(x)=1+x-\frac{x^2}{2}-\frac{x^3}{6}+\frac{x^4}{24}-\frac{x^5}{120}-\frac{x^6}{720}+\frac{(1)x^7}{5040}+\frac{(1)x^8}{40320}$$
$$+\frac{(-1)x^9}{362880}+\frac{(-0.999997)x^{10}}{3628800}+\frac{(-1)x^{11}}{39916800} \tag{12}$$

Example 2

For $x \in [0,1]$, the following linear boundary value problem is considered

$$\left.\begin{aligned} &y^{(13)}(x) = \ell^{-x} y^2 \\ &y^{(0)}(0) = 1, \\ &y^{(1)}(0) = 1, \\ &y^{(2)}(0) = 1, \\ &y^{(3)}(0) = 1, \\ &y^{(4)}(0) = 1, \\ &y^{(5)}(0) = 1, \\ &y^{(6)}(0) = 1, \\ &y^{(0)}(1) = \ell, \\ &y^{(1)}(1) = \ell, \\ &y^{(2)}(1) = \ell, \\ &y^{(3)}(1) = \ell, \\ &y^{(4)}(1) = \ell, \\ &y^{(5)}(1) = \ell, \end{aligned}\right\} \quad (13)$$

The analytical solution of the given problem from [18]

$$y(x) = \ell^{x} \quad (14)$$

The given thirteen order boundary value problem can be transformed with the following system

$$\left.\begin{aligned}
\frac{\mathrm{d}y}{\mathrm{d}x} &= p(x) \\
\frac{\mathrm{d}p}{\mathrm{d}x} &= q(x) \\
\frac{\mathrm{d}q}{\mathrm{d}x} &= r(x) \\
\frac{\mathrm{d}r}{\mathrm{d}x} &= s(x) \\
\frac{\mathrm{d}s}{\mathrm{d}x} &= j(x) \\
\frac{\mathrm{d}j}{\mathrm{d}x} &= u(x) \\
\frac{\mathrm{d}u}{\mathrm{d}x} &= z(x) \\
\frac{\mathrm{d}m}{\mathrm{d}x} &= n(x) \\
\frac{\mathrm{d}n}{\mathrm{d}x} &= l(x) \\
\frac{\mathrm{d}l}{\mathrm{d}x} &= g(x) \\
\frac{\mathrm{d}g}{\mathrm{d}x} &= v(x) \\
\frac{\mathrm{d}v}{\mathrm{d}x} &= y^2 \mathrm{e}^{-x}
\end{aligned}\right\} \qquad (15)$$

With

$$\begin{aligned}
&y(0)=1,\ p(0)=1,\ q(0)=1,\ r(0)=1,\ s(0)=1, \\
&j(0)=1,\ u(0)=1 \\
&z(0)=a,\ m(0)=b,\ n(0)=c,\ l(0)=d,\ g(0)=f, \\
&v(0)=h
\end{aligned}$$

This system of differential Equations (14) can be written in terms of the following system of integral equations with Lagrange multipliers $\lambda_i = +1,\ i = 1, 2, 3, \cdots, 13$

$$\left.\begin{aligned}
y^{k+1}(x) &= 1 + \int_0^x p^k(t)\,\mathrm{d}t \\
p^{k+1}(x) &= 1 + \int_0^x q^k(t)\,\mathrm{d}t \\
q^{k+1}(x) &= 1 + \int_0^x r^k(t)\,\mathrm{d}t \\
r^{k+1}(x) &= 1 + \int_0^x s^k(t)\,\mathrm{d}t \\
s^{k+1}(x) &= 1 + \int_0^x j^k(t)\,\mathrm{d}t \\
j^{k+1}(x) &= 1 + \int_0^x u^k(t)\,\mathrm{d}t \\
u^{k+1}(x) &= 1 + \int_0^x z^k(t)\,\mathrm{d}t \\
z^{k+1}(x) &= a + \int_0^x m^k(t)\,\mathrm{d}t \\
m^{k+1}(x) &= b + \int_0^x n^k(t)\,\mathrm{d}t \\
n^{k+1}(x) &= c + \int_0^x L^k(t)\,\mathrm{d}t \\
L^{k+1}(x) &= d + \int_0^x g^k(t)\,\mathrm{d}t \\
g^{k+1}(x) &= f + \int_0^x v^k(t)\,\mathrm{d}t \\
v^{k+1}(x) &= h + \int_0^x \left(y^k_{(t)}\right)^2 \mathrm{e}^{-x}\,\mathrm{d}t
\end{aligned}\right\} \quad (16)$$

With

$$\left.\begin{array}{l} y^0(x)=1, \\ p^0(x)=1 \\ q^0(x)=1 \\ r^0(x)=1 \\ s^0(x)=1 \\ j^0(x)=1 \\ u^0(x)=1 \\ z^0(x)=a \\ m^0(x)=b \\ n^0(x)=c \\ l^0(x)=d \\ g^0(x)=f \\ v^0(x)=h \end{array}\right\} \tag{17}$$

The above system of differential Equations (16) and (17) provide the following approximations

$$y^1(x)=1+\int_0^x p^0(x)\mathrm{d}x=1+\int_0^x(1)\mathrm{d}x=1+x$$
$$p^1(x)=1+\int_0^x \mathrm{d}x=1+x$$
$$q^1(x)=1+x$$
$$r^1(x)=1+x$$
$$s^1(x)=1+x$$
$$j^1(x)=1+x$$
$$u^1(x)=1+ax$$
$$z^1(x)=a+bx$$
$$m^1(x)=b+cx$$
$$n^1(x)=c+dx$$
$$L^1(x)=d+fx$$
$$g^1(x)=f+hx$$
$$v^1(x)=h+\int_0^x\left(y^0\right)^2\mathrm{e}^{-x}=h+\int_0^x(1)^2\mathrm{e}^{-x}$$
$$v^1=h+1-\mathrm{e}^{-x}$$

$k = 1$

$$y^2(x) = 1 + \int_0^x p^1(1)\,dx = 1 + \int_0^x (1+x)\,dx = 1 + x + \frac{x^2}{2}$$

$$\vdots$$

$$y^{(13)}(x) = 1 + x + \frac{x^2}{2} + \frac{x^3}{6} + \frac{ax^4}{24} + \frac{bx^5}{120} + \frac{cx^6}{720} + \frac{dx^7}{5040}$$
$$+ \frac{x^8}{40320} + \frac{x^9}{362880} + \frac{x^{10}}{3628800} + \frac{x^{11}}{39916800}$$
$$+ \frac{(-1+2a)x^{12}}{479001600} + \frac{(-1+2b)x^{13}}{6227020800}$$

The solution is given as

$$y(x) = 1 + x + \frac{x^2}{2} + \frac{x^3}{6} + \frac{ax^4}{24} + \frac{bx^5}{120} + \frac{cx^6}{720} + \frac{dx^7}{5040}$$
$$+ \frac{x^8}{40320} + \frac{x^9}{362880} + \frac{x^{10}}{3628800} + \frac{x^{11}}{39916800}$$
$$+ \frac{(-1+2a)x^{12}}{479001600} + \frac{(-1+2b)x^{13}}{6227020800} \quad (18)$$

The coefficients a, b, c and d were obtained using MAPLE 13 with the boundary condition at $x = 1$ given $a = 1, b = 1, c = 0.9999, d = 1.0001$.

The series solution can thus be written as

$$y(x) = 1 + x + \frac{x^2}{2} + \frac{x^3}{6} + \frac{x^4}{24} + \frac{x^5}{120} + \frac{0.9999x^6}{720} + \frac{1.00001x^7}{5040} + \frac{x^8}{40320} + \frac{x^9}{362880} + \frac{x^{10}}{3628800} + \frac{x^{11}}{39916800} + \frac{x^{12}}{479001600} + \frac{x^{13}}{6227020800} \tag{19}$$

5. RESULTS AND DISCUSSION

In example 1, Equation (13) is the numerical solution of Equation (7) using variational iteration technique. **Table 1** below shows the listof the results obtained by Variational Iteration Method and compared with the Equation (8) which is the exact solution of Equation (7). As we see from **Table 1**; it is clear that the result obtained by the present method is very superior to that obtained by the exact solution with highly accurate. Also the exact solution and variational iteration method solution are depicted in **Figure 1**. As we see, there is a very good agreement between the approximate solution obtained by the eleventh iteration using variational iteration method and the exact solution which is shown in **Table 1**.

In example 2, the comparison of the exact solution with the series solution of the Equation (20), obtained using the variational iteration technique is shown in **Table 2**.

The exact solution and the series solution obtained are depicted in **Figure 2**. There is a very good agreement and relationship between the series solution obtained by 13th iteration using variational iterations method and the exact solution which is shown in **Table 2**.

Table 1. Comparison of the exact solution with the series solution of the problem (7).

x	**Exact Solution**	**VIM**	**Error in Present Method**
0	1	1	0
0.1	0.994054	0.994054	3.88578E-15
0.2	0.931864	0.931864	1.46216E-13
0.3	0.769356	0.769356	8.80518E-13
0.4	0.784691	0.784691	2.35822E-12
0.5	0.659287	0.659287	3.8014E-12
0.6	0.537115	0.537115	5.14766E-11
0.7	0.381117	0.381117	1.56224E-11
0.8	0.240714	0.240714	8.99409E-11
0.9	0.129106	0.129106	4.70031E-10
1	0	0	2.06386E-9

The comparison of the exact solution with the series solution of the problem (7) obtained using the variational iteration technique.

6. CONCLUSIONS

In this paper, variational iteration method has been successfully employed to obtain the approximate solutions of boundary value problem for thirteenth-order differential equations. The method is applied in a direct way without using linearization, transformation, discretization, perturbation or restriction assumptions. We concluded that the proposed technique is very powerful and efficient in finding the analytical solution for wide classes of boundary value problems.

Table 2. Comparison of the exact solution with the series solution of the problem (14).

x	Exact Solution	VIM	Error in Present Method
0	1	1	0
0.1	1.105170918	1.105170918	4.17444E-14
0.2	1.221402758	1.221402758	2.64144E-12
0.3	1.349858808	1.349858808	2.99314E-11
0.4	1.491824698	1.491824697	1.67101E-10
0.5	1.648721271	1.64872127	6.30955E-10
0.6	1.8221188	1.822118799	1.84757E-09
0.7	2.013752707	2.013752703	4.47866E-09
0.8	2.225540928	2.225540919	9.21592E-09
0.9	2.459603111	2.459603095	1.58906E-08
1	2.718281828	2.718281808	2.09057E-08

The comparison of the exact solution with the series solution of the problem (14) obtained using variational iteration technique.

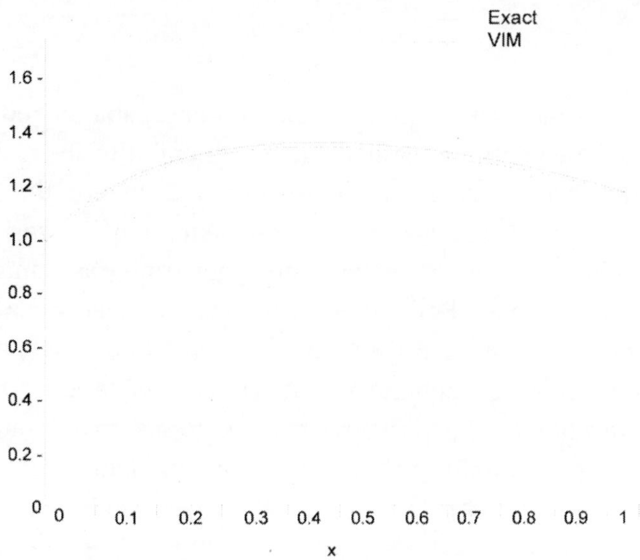

Figure 1. The graph show the exact and VIM solution for both Equation (8) and Equation (13) respectively.

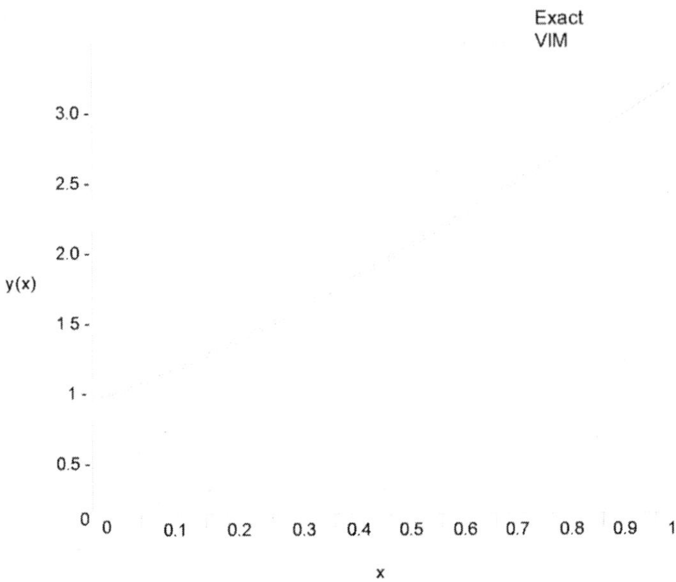

Figure 2. The graph show the exact and VIM solution for both Equation (15) and Equation (20) respectively.

It is worthwhile mentioning that the method is capable of reducing the volume of the computational work as compared to the classical methods. This can be seen in Tables 1 and 2, where the exact solution agreed with the series solution generated by variatinal iteration method. The error estimation is generated as compared with the exact method. Both approaches can be seen clearly on Figures 1 and 2. The method is extremely simple to be used, and is accurate for solving nonlinear differential equation. Hence variational iteration method is a powerful tool to search for solutions of various linear and nonlinear boundary value problems.

REFERENCES

1. A. M. Wazwaz, "The Variational Iteration Method: A Reliable Analytic Tool for Solving Linear and Nonlinear Wave Equations," Computer and Mathematics with Applications, Vol. 54, No. 7-8, 2007, pp. 926-932.

2. M. Inc and D. J. Evans, "An Efficient Approach to Approximate Solutions of Eight-Order Boundary-Value Problems," International Journal of Computer Mathematics, Vol. 81, No. 6, 2007, pp. 685-692.

3. R. P. Agarwal, "Boundary Value Problems for HigherOrder Differential Equations," World Scientific, Singapore, 1986.

4. S. Chandrasekhar, "Hydrodynamic Hydromagnetic Stability," Clarendon Press, Oxford, New York, 1981.

5. A Boutayeb E. H. Twizell and K. Djidjeli, "Numerical Methods for Eight-, Tenthand Twelfth-Order Eigenvalue Problems Arising in Thermal Instability," Advances in Computational Mathematics, Vol. 2, No. 4, 1994, pp. 407- 436.

6. J. H. He, "Approximate Analytical Solution for Seepage Flow with Fractional Derivatives in Porous Media," Computer Methods in Applied Mechanics and Engineering, Vol. Vol. 167, No. 1-2, 1998, pp. 57-68.

7. S. Elham, "Variational Iteration Method—A Kind of Nonlinear Analytical Technique: Some Examples," International Journal of Non-Linear Mechanics, Vol. 34, No. 4, 1999, pp. 699-708.

8. M. Yulita, "Variational Method for Outonomous Ordinary Differential Equations," Applied Mathematics and Computation, Vol. 114, No. 2-3, 2000, pp. 115-123.

9. A. S. Arife, "Variational Theory for Linear Magneto— Elasticity," International Journal of Nonlinear Sciences and Numerical Simulation, Vol. 2, No. 4, 2001, pp. 309- 316.

10. L. M. Watson and M. R. Scott, "Solving Spline-Collocation Approximations to Nonlinear Two-Point Boundary Value Problems by a Homotopy Method," Applied Mathematics and Computation, Vol. 24, No. 4, 1987, pp. 333-357.

11. G. R. Liu and T. Y. Wu, "Differential Quadrature Solutions of Eight-Order Boundary-Value Differential Equations," Journal of Computational and Applied Mathematics, Vol. 145, No. 1, 1973, pp. 223-235.

12. T. Mura, M. Inokuti and H. Sekine, "General Use of Lagrange Multiplier Innonlinear Mathematical Physics," In: Variational Method in the Mech. of Solids, Pergamon Press, New York, 1978, pp. 156-162.

13. H. A. Watts and M. R. Scott, "A Systematical Collection of Codes for Solving Two-Point BVPs, Numerical Methods for Differential Systems," Academic Press, Waltham, 1976.

14. M. R. Scott and H. A. Watts, "Computational Solutions of Linear Two-Point Boundary Value Problems via Orthonormalization," SIAM Journal on Numerical Analysis, Vol. 14, No. 1, 1977, pp. 40-70.

15. G. Akram, S. S. Siddiqi and S. Zaheer, "Solution of Eighth Order Boundary Value Problems Using Variational Iteration Technique," Submitted in European Journal of Scientific Research, 2007.

16. S. S. Siddiqi and G Akram, "Solutions of Tenth-Order Boundary Value Problems Using Eleventh Degree Spline," Applied

Mathematics and Computation, Vol. 185, No. 1, 2007, pp. 115-127.

17. S.S Siddiqi and S. Zaheer, "Solutions of Tenth-Order Boundary Value Problems Using Non-Polynomial Spline Technique," Applied Mathematics and Computation, Vol. 190, 1977, pp. 641-651.

18. S. S. Siddiqi and E. H. Twizell, "Spline Solution of Linear Tenth-Order Boundary Value Problems," International Journal of Computer Mathematics, Vol. 68, No. 3-4, 1998, pp. 345-362.

CHAPTER 3

A Five-Step P-Stable Method for the Numerical Integration of Third Order Ordinary Differential Equations

D. O. Awoyemi[1], S. J. Kayode[2], L. O. Adoghe[3]

[1]Department of Mathematics, Landmark University, Umuaru, Nigeria

[2]Department of Mathematical Sciences, Federal University of Technology, Akure, Nigeria

[3]Department of Mathematics, Ambrose Alli University, Ekpoma, Nigeria

ABSTRACT

In this paper we derived a continuous linear multistep method (LMM) with step number k = 5 through collocation and interpolation techniques using power series as basis function for approximate solution. An order nine p-stable scheme is developed which was used to solve the third order initial value problems in ordinary differential equation without first reducing to a system of first order equations. Taylor's series algorithm of the same order was developed to implement our method. The result obtained compared favourably with existing methods.

KEYWORDS

continuous Collocation, Multistep Methods, Interpolation, Third Order, Power Series, approximate Solution

1. INTRODUCTION

Linear multistep methods (LMM) for solving first order initial value problems (ivps) is of the form

$$\sum_{j=0}^{k}\alpha_j y_{n+j} = h\sum_{j=0}^{k}\beta_j f_{n+j} \tag{1}$$

where α_j and β_j are uniquely determined and $\alpha_0 + \beta_j \neq 0, \alpha_k = 1$

Conventionally, they are used to solve higher order ordinary differential equations by first reducing them to a system of first order. This approach has been extensively discussed in [1] -[5] . However the method of reducing to a system first order has some serious drawback which includes wastage of human effort and computer time [6] .

The LMM in (1) generates discrete schemes which are used to solve first order odes. Various forms of this LMM have been developed [1] -[4] . Other researchers have introduced the continuous LMM using the continuous collocation and interpolation technique. This has led to the development of continuous LMM of form

$$Y(x)=\sum_{j=0}^{k}\alpha_j(t)y_{n+j}+h\sum_{j=0}^{k}\beta_j(t)f_{n+j} \quad (2)$$

α_j and β_j are expressed as continuous functions of t and are at least differentiable once.

The introduction of continuous collocation methods as against the discrete schemes enhances better global error estimation and ability to approximate solution at all interior points [6] - [10] . In this study, we shall develop continuous multistep collocation method for the solution of third order ordinary differential equations using power series as the basis function.

1.1 Power Series Collocation

In [6] [8] [9] , some continuous LMM of Type (2) were developed using power series of form

$$Y(x)=\sum_{j=0}^{k}a_jx^j \quad (3)$$

In [10] Chebyshev polynomial function of the form

$$Y(x)=\sum_{j=0}^{k}a_jT_j(x)\left(\frac{x-x_k}{h}\right) \tag{4}$$

where $T_j(x)$ are some Chebyshev function used to develop continuous LMM.

The use power series as basis function for derivation of continuous LMM are based on the property of analytic function that given the Taylor's polynomial of the form

$$P(x)=\sum_{r=0}^{n}\frac{y^r(x_0)(x-x_n)}{r!} \tag{5}$$

The approximate function $p(x)$ reduces to $Y(x)$ as $n\to\infty$

In this study we proposed the polynomial function of the form in [7] :

$$Y(x)=\sum_{j=0}^{k}a_j(x-x_n)^j \tag{6}$$

which is of Type (3) to develop a continuous LMM for the solution of initial value problem of the form:

$$y'''=f(x,y,y',y''),y(a)=y_0; \quad y'(a)=\partial_0,y''(a)=\partial_N \tag{7}$$

This paper is organized as follows: Section 1 consists of introduction and background of study; Section 2, we derive a continuous approximation to $Y(x)$ for exact solution $y(x)$, and specific methods;

section 3 consists of the analysis and implementation followed by numerical examples.

2. DERIVATION OF THE METHOD

Consider the third order differential Equation (7), we proposed an approximate solution of the form:

$$y_k(x) = \sum_{j=0}^{M} a_j \psi_j(x) \tag{8}$$

where $\psi_j(x) = (x - x_n)^j, j \geq 0$.

The derivative of (8) up to the third order yield

$$y_k'''(x) = \sum_{j=0}^{k} j(j-1)(j-2) a_j \psi_{j-3}(x) \tag{9}$$

And $a \leq x \leq b$ where $a_j's$ are the parameters to be determined. By substituting (8) and (9) into (7) we have

$$\sum_{j=0}^{k} j(j-1)(j-2) a_j \psi_{j-3}(x) = f(x, y, y', y'') \tag{10}$$

Collocating (10) at $x = x_{n+i}, i = 0(1)m$ and interpolating (8) at $x = x_{n+i}, i = 0(1)m-1$ we obtained the system of equations given below

$$\sum_{j=0}^{M} a_j \psi_j (x_{n+i}) = y_{n+i} \tag{11}$$

$$\sum_{j=0}^{M} j(j-1)(j-2) a_j \psi_{j-3}(x) = f_{n+i} \tag{12}$$

The above equations are solved to obtain the values of $a_j's$ which when substituted into Equation (6) yield a method of the form in Equation (13)

$$Y(x) = \sum_{j=0}^{k-1} \alpha_j(t) y_{n+j} + h \sum_{j=0}^{k} \beta_j(t) f_{n+j} \tag{13}$$

The continuous polynomial obtained when the values of $a_j's$ are substituted into (6) and simplified is as follows

$$
\begin{aligned}
Y(x)=&\frac{1}{59520h^{10}}\Big[59520h^{10}-3305158h^{9}(x-x_n)+5968895h^{8}(x-x_n)^{2}-8146800h^{6}(x-x_n)^{4}\\
&+8869224h^{5}(x-x_n)^{5}-4685898h^{4}(x-x_n)^{6}+1424760h^{3}(x-x_n)h^{7}-253425h^{2}(x-x_n)^{8}\\
&+24490h(x-x_n)^{9}-1024(x-x_n)^{10}\Big]y_n\\
&+\frac{1}{59520h^{10}}\Big[6211314h^{9}(x-x_n)-11400365h^{8}(x-x_n)^{2}+156277600h^{6}(x-x_n)^{4}\\
&-17084088h^{5}(x-x_n)^{5}+9069483h^{4}(x-x_n)^{6}-2770920h^{3}(x-x_n)^{7}+495075h^{2}(x-x_n)^{8}\\
&-48030h(x-x_n)^{9}+1952(x-x_n)^{10}\Big]y_{n+1}\\
&+\frac{1}{59520h^{10}}\Big[898062h^{9}(x-x_n)-158251h^{8}(x-x_n)^{2}+199800h^{6}(x-x_n)^{4}-1963080h^{5}(x-x_n)^{5}\\
&+907074h^{4}(x-x_n)^{6}-235800h^{3}(x-x_n)^{7}+35325h^{2}(x-x_n)^{8}-2850h(x-x_n)^{9}+96(x-x_n)^{10}\Big]y_{n+2}\\
&+\frac{1}{59520h^{10}}\Big[-7311962h^{9}(x-x_n)+13490545h^{8}(x-x_n)^{2}-18291600h^{6}(x-x_n)^{4}\\
&+19701528h^{5}(x-x_n)^{5}-10278870h^{4}(x-x_n)^{6}+3085320h^{3}(x-x_n)^{7}-542175h^{2}(x-x_n)^{8}\\
&+51830h(x-x_n)^{9}-2080(x-x_n)^{10}\Big]y_{n+3}\\
&+\frac{1}{399974400h^{7}}\Big[-160164924h^{9}(x-x_n)+27846271h^{8}(x-x_n)^{2}+6666240h^{7}(x-x_n)^{3}\\
&-50359720h^{6}(x-x_n)^{4}+518336784h^{5}(x-x_n)^{5}-269004260h^{4}(x-x_n)^{6}+81125040h^{3}(x-x_n)^{7}\\
&-14366570h^{2}(x-x_n)^{8}+98910h(x-x_n)^{9}\Big]-56000(x-x_n)^{10}\Big]f_n\\
&+\frac{1}{399974400h^{7}}\Big[-9863776238h^{9}(x-x_n)+18418906560h^{8}(x-x_n)^{2}-25437064780h^{6}(x-x_n)^{4}\\
&+27743628290h^{5}(x-x_n)^{5}-14672416610h^{4}(x-x_n)^{6}-4463635120h^{3}(x-x_n)^{7}\\
&-794196650h^{2}(x-x_n)^{8}+5482910h(x-x_n)^{9}-3109568(x-x_n)^{10}\Big]f_{n+1}\\
&+\frac{1}{399974400h^{7}}\Big[-2355609311h^{9}(x-x_n)+43809648630h^{8}(x-x_n)^{2}-59823556780h^{6}(x-x_n)^{4}\\
&+64857656850h^{5}(x-x_n)^{5}-34104473030h^{4}(x-x_n)^{6}+10321466720h^{3}(x-x_n)^{7}\\
&+1828259700h^{2}(x-x_n)^{8}+12572860h(x-x_n)^{9}-7107968(x-x_n)^{10}\Big]f_{n+2}\\
&+\frac{1}{399974400h^{7}}\Big[-10969770700h^{9}(x-x_n)+20392494160h^{8}(x-x_n)^{2}-27749220780h^{6}(x-x_n)^{4}\\
&+29987527570h^{5}(x-x_n)^{5}-157505754790h^{4}(x-x_n)^{6}+4732439520h^{5}(x-x_n)^{5}\\
&+834535300h^{2}(x-x_n)^{8}+5715660h(x-x_n)^{9}-3219328(x-x_n)^{10}\Big]f_{n+3}\\
&+\frac{1}{399974400h^{7}}\Big[-177230988h^{9}(x-x_n)+329512510h^{8}(x-x_n)^{2}\\
&-449080800h^{6}(x-x_n)^{4}+486136784h^{5}(x-x_n)^{5}-255405332h^{4}(x-x_n)^{6}\\
&+77358320h^{3}(x-x_n)^{7}+13746850h^{2}(x-x_n)^{8}+95110h(x-x_n)^{9}-54208(x-x_n)^{10}\Big]f_{n+4}\\
&+\frac{1}{399974400h^{7}}\Big[-2399724h^{9}(x-x_n)+4458670h^{8}(x-x_n)^{2}-5990880h^{6}(x-x_n)^{4}+6383440h^{5}(x-x_n)^{5}\\
&-3620852h^{4}(x-x_n)^{6}+941680h^{3}(x-x_n)^{7}-154930h^{2}(x-x_n)^{8}+950h(x-x_n)^{9}-448(x-x_n)^{10}\Big]f_{n+5}
\end{aligned}
\tag{14}
$$

Evaluating (14) at $x = x_{n+5}$ the following discrete method is obtained

$$y_{n+5} - \frac{29}{31}y_{n+4} - \frac{68}{31}y_{n+3} + \frac{68}{31}y_{n+2} + \frac{29}{31}y_{n+1} - y_n$$
$$= h^3\left(\frac{21}{2480}f_{n+5} + \frac{1177}{2480}f_{n+4} + \frac{1921}{1240}f_{n+3} + \frac{1921}{1240}f_{n+2} + \frac{1177}{2480}f_{n+1} + \frac{21}{2480}f_n\right)$$

(15)

3. ANALYSIS AND IMPLEMENTATION OF THE METHOD

3.1. Basic properties of the Method

The method (15) is a specific member of the conventional LMM which can expressed as

$$\sum_{j=0}^{k} \alpha_j y_{n+j} = h^3 \sum_{j=0}^{k} \beta_j y'''_{n+j}$$

(16)

Following [1] [2] , we define the local truncation error associated with (16) by the difference operator

$$L[y(x);h] = \sum_{j=0}^{k}\left[\alpha_j y(x_n + jh) - h^3 \beta_j y'''(x_n + jh)\right]$$

(17)

where $y(x)$ is assumed to have continuous derivatives of sufficiently high order. Therefore expanding (23) in Taylor series about the point x to obtain the expression

$$L[y(x);h] = C_0 y(x) + C_1 hy'(x) + C_2 hy''(x) + \cdots + C_p h^p y^p(x) + C_{p+1} h^{p+1} y^{p+1}(x) + C_{p+2} h^{P+2} y^{p+2}(x) \quad (18)$$

where the $C_0, C_1, C_2, \cdots, C_{p+2}$ are defined as

$$C_0 = \sum_{j=0}^{k} \alpha_j$$

, , ,

In the sense of [1] , we say that the method (20) is of order p and error constant $C_{p+2} \neq 0$ if

$$C_0 = C_1 = C_2 = \cdots = C_{p+1} = 0, C_{p+2} \neq 0$$

Using the concept above, the method (19) has order $p = 9$ and error constant given by

$$C_{p+2} = -0.005403 C_{p+2}$$

3.2. Zero-stability of the 5-Step Method

Considering the first characteristics polynomial of the method of Equation (15) given as

$$\rho(r)=r^5-\frac{29}{31}r^4-\frac{68}{31}r^3+\frac{68}{31}r^2+\frac{29}{31}r-1=0$$

Putting $r=1,\rho(r)=0$ implying that $r-1$ is a factor. Therefore solving the polynomial it is found that $(r-1)^3$ is also a factor of the polynomial and $r=1,1,1$. The other roots which are called spurious roots are $r=-0.776216472$ and -1.28829956.

3.3. Region of absolute Stability of the 5-Step scheme

Applying the boundary locus method, we have that

$$\rho(r)=r^5-\frac{29}{31}r^4-\frac{68}{31}r^3+\frac{68}{31}r^2+\frac{29}{31}r-1=0$$

$$\sigma(r)=\frac{h^3}{2480}\left(21r^5+1177r^4+3842r^3+3842r^2\sigma(r)+1177r+21\right)$$

In the spirit of Lambert (1973),

$$h(r)=\frac{\rho(r)}{\sigma(r)}=\frac{\rho(\mathrm{e}^{\mathrm{i}\theta})}{\sigma(\mathrm{e}^{\mathrm{i}\theta})}$$

By letting $r=\mathrm{e}^{\mathrm{i}\theta}=\cos\theta+\mathrm{i}\sin\theta$ and substituting this into the express above to yield

$$x(\theta)=\frac{2480\{\cos 5\theta-29\cos 4\theta-68\cos 3\theta+68\cos 2\theta+29\cos\theta+1\}}{21\cos 5\theta+1177\cos 4\theta+3842\cos 3\theta+3842\cos 2\theta+1177\cos\theta}$$

At $\theta=0^{\circ}$ and $\theta=180^{\circ}$ for $0^{\circ}\leq\theta\leq 180^{\circ}$ at an interval of 30° we have that $X(\theta)=(0,\infty)$. The method is therefore said to be p-stable.

3.4. Implementattion

Single step method can be used to solve higher order ordinary differential equations directly without the need to first reducing it to an equivalent system of first order.

Consider the initial value problem in (7). For our method of order $p=9$, Taylor series expansion is used to calculate.

$y_{n+1}, y_{n+2}, y_{n+3}$ and their first, second, third derivatives up to order $p=9$.

$$\begin{aligned} y_{n+j} &\equiv y(x_n+jh) \\ &\cong y(x_n)+jhy'(x_n)+\frac{(jh)^2 y''(x_n)}{2!}+\frac{(jh)^3 f_n}{3!}+\frac{(jh)^4 f_n'}{4!} \\ &\quad+\frac{(jh)^5 f_n''}{5!}+\frac{(jh)^6 f_n'''}{6!}+\frac{(jh)^7 f_n^{iv}}{7!}+\frac{(jh)^8 f_n^{v}}{8!}+\frac{(jh)^9 f_n^{vii}}{9!} \end{aligned}$$

$j=1(1)4$

$$y'_{n+j} \equiv y'(x_n + jh) \cong y'(x_n) + jhy''(x_n) + \frac{(jh)^2 f_n}{2!} + \frac{(jh)^3 f'_n}{3!} + \frac{(jh)^4 f''_n}{4!} + \frac{(jh)^5 f'''_n}{5!} + \frac{(jh)^6 f^{iv}_n}{6!}$$
$$+ \frac{(jh)^7 f^{v}_n}{7!} + \frac{(jh)^8 f^{vi}_n}{8!} + \frac{(jh)^9 f^{vii}_n}{9!} + \cdots$$

$$y''_{n+j} \equiv y''(x_n + jh)$$
$$\cong y''(x_n) + jhf_n + \frac{(jh)^2 f'_n}{2!} + \frac{(jh)^3 f''_n}{3!}$$
$$+ \frac{(jh)^4 f'''_n}{4!} + \frac{(jh)^5 f^{iv}_n}{5!} + \frac{(jh)^6 f^{v}_n}{6!} + \frac{(jh)^7 f^{vi}_n}{7!} + \frac{(jh)^8 f^{vii}_n}{8!} + \frac{(jh)^9 f^{viii}_n}{9!}$$

$$y'''_{n+j} \equiv y'''(x_n + jh)$$
$$\cong f_n + jhf'_n + \frac{(jh)^2 f''_n}{2!} + \frac{(jh)^3 f'''_n}{3!} + \frac{(jh)^4 f^{iv}_n}{4!}$$
$$+ \frac{(jh)^5 f^{v}_n}{5!} + \frac{(jh)^6 f^{vi}_n}{6!} + \frac{(jh)^7 f^{vii}_n}{7!} + \frac{(jh)^8 f^{viii}_n}{8!} + \frac{(jh)^9 f^{ix}_n}{9!} + \cdots$$

Then the known values of x_n and y_n are substituted into the differential equations. Next the differential equation is differentiated to obtain the expression for higher derivatives using partial differentiation as follows

$$y''' = f(x, y, y', y'') = f_j$$

$$y^{iv} = f_x + y'f_y + y''f_{y'} + ff_{y''} = \left(\frac{\partial}{\partial x} + y'\frac{\partial}{\partial y} + y''\frac{\partial}{\partial y'} + f\frac{\partial}{\partial y''}\right) = Df_j$$

$$y^{v} = f_{xx} + (y')^2 f_{yy} + (y'')^2 f_{y'y'} + f^2 f_{y'y''} + 2y'ff_{xy} + 2y''f_{xy'} + 2ff_{xy''}$$
$$+ 2y'y''f_{yy'} + 2y'ff_{yy''} + 2y''ff_{y'y''} + Df_j(f_{y''}) + f_j(y'' + f_{y'})$$
$$= D^2 f_j + (f_{y''})Df_j + f_j(y'' + f_{y'})_j = D^2 f_j + (f_{y''})Df_j + f_j(y'' + f_{y'})_j$$

Where

$$D=\left(\frac{\partial}{\partial x}+y'\frac{\partial}{\partial y}+y''\frac{\partial}{\partial y'}+f\frac{\partial}{\partial y''}\right)$$ and

$D^P f_j$ where p is the order of the method.

3.5. Numerical Experiments

Our methods of order $p=9$ were used to solve some initial value problems of both general and special nature using Taylor's series. Our results were compared with the results of other researchers in this area as seen in table 1. In table 2 and table 3, the accuracy of our method is seen in the small error values.

The following initial value problems were used as our test problems:

3.6. Problem 1

$$y'''=-y, y(0)=1, y'(0)=-1, y''(0)=1.$$
$$o\le x\le 1, h=0.1$$

Exact solution: $y(x)=\mathrm{e}^{-x}$.

3.7. Problem 2

$$.y''' - 2y'' - 3y' + 10y = 34xe^{-2x} - 16e^{-2x} - 10x^2 + 6x + 34.$$

$$y(0) = 3, y'(0) = 0, y''(0) = 0, 0 \le x \le 1$$

Exact solution: $y(x) = x^2 e^{-2x} - x^2 + 3$.

3.8. Problem 3

$$y''' = e^x$$
,

Exact solution: $2 + 2x^2 + e^x$.

Table 1. showing the result of test problem 1.

X	Exact solution	New result method (P = 9)	Error in our for (P = 9)	Error in [11] (P = 9)
0.1	0.904837	0.940837	0.0000+00	2.1760E−12
0.2	0.818731	0.818731	2.7756E−14	1.3935E−11
0.3	0.740818	0.740818	1.5838E−12	3.4443E−11
0.4	0.670320	0.670320	2.7879E−11	6.4477E−11
0.5	0.606531	0.606531	2.9477E−11	1.0316E−10
0.6	0.548812	0.548812	8.5048E−11	1.4979E−10
0.7	0.496585	0.496585	8.0357E−11	2.0486E−10
0.8	0.449329	0.449329	1.6601E−10	2.6756E−10
0.9	0.406570	0.406570	1.1176E−10	6.9382E−10
1.0	0.367879	0.367879	1.4871E−10	1.4224E−10

Table 2. Showing the result of test problem 2.

X	Exact solution	New result method (P = 9)	Error in our for (P = 9)
0.1	0.299819E+01	0.299819E+01	6.6218E−13
0.2	0.298681E+1	0.298681E+01	6.2238E−11
0.3	0.295939E+01	0.295939E+01	3.5134E−09
0.4	0.291189E+01	0.291189E+01	6.1100E−07
0.5	0.284197E+01	0.284197E+01	6.4183E−07
0.6	0.274843E+01	0.274843E+01	1.8082E−06
0.7	0.263083E+01	0.263083E+01	1.3511E−06
0.8	0.248921E+01	0.248921E+01	1.3367E−06
0.9	0.232389E+01	0.232390E+01	7.9041E−06
1.0	0.213534E+01	0.213537E+01	3.7360E−05

Table 3. showing the result of test problem 3.

X	Exact solution	New result method (P = 9)	Error in our for (P = 9)
0.1	0.321517E+01	0.321517E+01	0.0000E+00
0.2	0.330140E+01	0.330140E+01	2.8422E−13
0.3	0.352980E+01	0.352980E+01	1.6729E−12
0.4	0.381182E+01	0.381182E+01	2.9983E−11
0.5	0.414872E+01	0.414872E+01	3.1673E−11
0.6	0.454212E+01	0.454212E+01	9.1899E−11
0.7	0.499375E+01	0.499375E+01	8.9531E−11
0.8	0.550554E+01	0.550554E+01	1.9168E−10
0.9	0.607960E+01	0.607960E+01	2.1110E−10
1.0	0.671828E+01	0.671828E+01	4.9398E−10
1.1	0.742417E+01	0.742417E+01	8.6728E−10
1.2	0.820012E+01	0.820012E+01	2.3764E−09

4. DISCUSSION OF RESULT

We have developed and implemented our methods using Taylor series of the same order as the schemes that we developed. Some special and general third order initial value problems (ivps) were used to test the efficiency of our methods. Our method was found to be zero stable, consistent and convergent. The better accuracy of our method can be shown from the numerical examples.

REFERENCES

1. Lambert, J.D. (1973) Computional Methods in ODEs. John Wiley & Sons, New York.

2. Fatunla, S.O. (1988) Numerical Methods for Initial Value Problems in Ordinary Differential Equations. Academic Press Inc., Harcourt Brace, Jovanovich Publishers, New York.

3. Butcher, S.C. (2003) Numerical Methods for Ordinary Differential Equations. John Wiley & Sons, New York.

4. Henrici, P. (1962) Discrete Variable Method in Ordinary Differential Equations. John Wiley & Sons, New York.

5. Bruguano, L. and Trigiante. D. (1998) Solving Differential Problems by Multistep Initial and Boundary Value Methods. Gordon and Breach Science Publishers, Amsterdam, 280-299.

6. Awoyemi, D.O. (2003) A P-Stable Linear Multistep Method for Solving General Third Order Ordinary Differential Equations. International Journal of computer Mathematics, 80, 987-993.

7. Okunuga, S.A. and Ehijie, J. (2009) New Derivation of Continuous Multistep Methods Using Power Series as Basis Function. Journal of Modern Mathematics and Statistics, 3, 43-55.

8. Adesanya, A.O. (2011) Block Methods for Direct Solutions of General Higher Order Initial Value Problems of Ordinary Differential Equations. Ph.D. Thesis, the Department of Mathematical Sciences, Federal University of Technology, Akure.

9. Adeniyi, R.B. and Alabi, M.O. (2006) Derivation of Continuous Multistep Methods Using Chebyshev Polynomial Basis Functions. Abacus, 33, 351-361.

10. Onumanyi, P., Oladele, J.O., Adeniyi, R.B. and Awoyemi, D.O. (1993) Derivation of Finite Difference Method by Collocation. Abacus, 23, 2-83.

11. Olabode, B.T (2007) Some Linear Multistep Methods for Special and General Third Order Initial Value Problems in Ordinary Differential Equations. Ph.D. Thesis, the Department of Mathematical Sciences, The Federal University of Technology, Akure.

CHAPTER 4

Multiple Solutions to Implicit Symmetric Boundary Value Problems for Second Order Ordinary Differential Equations (ODEs): Equivariant Degree Approach

Zalman Balanov *, Wieslaw Krawcewicz, Zhichao Li and Mylinh Nguyen

Department of Mathematical Sciences, University of Texas at Dallas, Richardson, TX 75081, USA(M.N.)

ABSTRACT

In this paper, we develop a general framework for studying Dirichlet Boundary Value Problems (BVP) for second order symmetric implicit differential systems satisfying the Hartman-

Nagumo conditions, as well as a certain non-expandability condition. The main result, obtained by means of the equivariant degree theory, establishes the existence of multiple solutions together with a complete description of their symmetric properties. The abstract result is supported by a concrete example of an implicit system respecting D4-symmetries.

KEYWORDS

symmetric BVP; second order implicit Ordinary Differential Equation (ODE); multiple solutions; *a priori*bounds; equivariant degree; multivalued maps; dihedral group symmetries

1. INTRODUCTION

1.1. Subject and Goal

Boundary value/periodic problems for second order nonlinear Ordinary Differential Equations (ODEs) have been within the focus of the nonlinear analysis community for a long time (see, for example [1,2,3]). In [2,4], P. Hartman established the existence result for the boundary value problem:

$$\begin{cases} \ddot{y} = f(t, y, \dot{y}) \\ y(0) = y(1) = 0 \end{cases} \tag{1}$$

where the function, $f : [0,1] \times \mathbb{R}^n \times \mathbb{R}^n \to \mathbb{R}^n$, satisfies the so-called Hartman-Nagumo conditions, which, informally, means that *f* is a reasonable function having a sub-quadratic growth on $\dot{y}$.

Later on, H.W. Knobloch [5,6] observed that a similar result is true for the corresponding periodic problem:

$$\begin{cases} \ddot{y} = f(t, y, \dot{y}) \\ y(0) = y(1), \quad \dot{y}(0) = \dot{y}(1) \end{cases} \tag{2}$$

Several extensions of Hartman-Knobloch results of perturbations of the ordinary vector p-Laplacian operator were suggested by J. Mawhin *et al.* (see, for example, [7,8,9] and the references therein).

Although Hartman's *existence* result was extended to more general settings by many authors, to the best of our knowledge, the problems of *estimating a minimal number of solutions to* (1), as well as *classifying their symmetric properties* have not been carefully studied. To some extent, our recent paper [10] opened a door to a systematic usage of the *equivariant degree theory* for analysis of *multiple* solutions to *symmetric* (1) and its generalizations. The starting point for our discussion was Example 6.1 from [11] in which a particular case of BVP (1) in the presence of D4-symmetries was considered (see also [12] in which a "multivalued perturbation of this example" was discussed).

The *goal* of this paper is to study multiple solutions to boundary value problems for *implicit symmetric second order differential systems* using the equivariant degree-based method. To simplify our exposition, we will restrict ourselves to the Dirichlet boundary conditions. More specifically, we are interested in the BVPs of the form:

$$\begin{cases} \ddot{y} = h(t, y, \dot{y}, \ddot{y}), & \text{a.e. } t \in [0,1] \\ y(0) = y(1) = 0 \end{cases} \tag{3}$$

where V:=Rn is an orthogonal *G*-representation and h:[0,1]×V×V×V→V is a *G*-equivariant map satisfying the so-called Carathéodory condition. For some motivating examples from mechanics (including, in particular, the ones modeled by the so-called generalized Liénard equation), we refer the reader to [13] and the references therein.

1.2. Method

The main idea behind the method allowing us to study (3) can be traced back to [14,15]. Namely, assume that *h* satisfies the Hartman-Nagumo conditions with respect to (y,y˙), and, in addition, it is non-expansive with respect to y¨. Since the set of fixed points of a non-expansive map of an Euclidean space is *convex*, one can canonically associate with problem (3) the "explicit" differential *inclusion* of the form:

$$\begin{cases} \ddot{y} \in \tilde{F}(t, y, \dot{y}), & \text{a.e. } t \in [0,1] \\ y(0) = y(1) = 0 \end{cases} \tag{4}$$

with $\tilde{F} : [0,1] \times V \times V \to \mathcal{K}_c(V)$ (here, $\mathcal{K}_c(V)$ stands for the set of all non-empty convex and compact subsets of *V*). By the (equivariant) homotopy argument, the later problem can be reduced to the "explicit" *single-valued* symmetric BVP of the form of Equation (1), and the equivariant degree-based method developed in [10] can be applied.

Recall that the equivariant degree is a topological tool allowing "counting" orbits of solutions to (symmetric) equations in the same way as the usual Brouwer degree does, but according to their symmetric properties. This method is an alternative and/or complement to the equivariant singularity theory developed by M. Golubitsky *et al.* (see, for example, [16]), as well as to a variety of methods rooted in Morse Theory, Lusternik-Schnirelmann Theory and Morse-Floer complex techniques (see, for example, [17,18,19,20]) used for the treatment of variational problems with symmetries. These standard methods, although being quite effective in the settings in which they are usually applied, encounter technical difficulties when: (i) the group of symmetries is large; (ii) multiplicities of eigenvalues of linearizations are large; (iii) phase spaces are of a high dimension; and (iv) the operators involved exhibit a lack of smoothness. Furthermore, one would expect to use computer routines for complex computations, while it is not clear if these approaches are "open enough" to be computerized. On the other hand, the equivariant degree theory has all the attributes allowing its application in settings related to (i)–(iv), and in many cases, it allows computerization. For instance, in the case of the dihedral group, the tools required for the symbolic computations of the equivariant degree can be found at [21]. For a detailed exposition of the equivariant degree theory, we refer the reader to [11,22,23,24,25].

1.3. Overview

After the Introduction, the paper is organized as follows. In Section 2, we collect the standard equivariant background together with basic properties of the equivariant degree (without free parameters) for compact equivariant multivalued fields. This theory is applied in Section 3 for studying "explicit" second order equivariant inclusions (see (11)). We reformulate (11) as a fixed-point problem with a compact equivariant multivalued vector field defined on the Sobolev space, H2([0, 1];Rn), and associate to this field an invariant, ω(C,F), expressed in terms of the equivariant degree (see (22)). Using ω(C,F), we formulate our result for (11) (see Theorem 3.6). In Section 4, we combine Theorem 3.6 with the equivariant version of the well-known result from [15] (*cf.* Lemma 4.1) to obtain our main abstract result for the "explicit" BVP (3) (in fact, this result (see Theorem 4.4) is expressed in terms of ω(C,F) provided by Lemma 4.1). In Section 5, we describe a wide class of BVPs (3) symmetric with respect to the *dihedral* group representations for which Theorem 4.4 can be applied to obtain a complete symmetric classification of solutions (see Proposition 5.1). We also give a concrete D4-symmetric example supporting Proposition 5.1 (see Theorem 5.2). To make our exposition self-contained, we conclude with two Appendices (Appendix 1 is related to the equivariant degree theory for *single-valued* maps (in particular, the concept of a *Burnside ring*, and a computational formula for the equivariant degree of a linear equivariant isomorphism is given); inAppendix 2, we collected all the facts frequently used in this paper that are related to D4-representations and D4-equivariant degree).

2. *G*-ACTIONS AND EQUIVARIANT DEGREE WITHOUT PARAMETERS FOR MULTIVALUED FIELDS

In this section, we briefly recall the standard "equivariant jargon" and present basic facts related to the equivariant degree without free parameters for equivariant multivalued fields. In what follows, G stands for a *finite* group and V for an orthogonal G-representation.

2.1. G-Actions

For a subgroup, $H \subset G$, denote by $N(H)$ the normalizer of H in G, by $W(H) = N(H)/H$, the Weyl group of H in G, and by (H), the conjugacy class of H in G. The set, $\Phi(G)$, of all conjugacy classes in G admits a partial order defined as follows: $(H) \leq (K)$ if and only if $gHg^{-1} \subset K$ for some $g \in G$.

For a G-space, X and $x \in X$, denote by $G_X := \{g \in G : gx = x\}$ the *isotropy* of x and by $G(x) := \{gx : g \in G\} \simeq G/G_X$ the *orbit* of x. Given an isotropy, G_X, call (G_X) the *orbit type* in X and put $\Phi(G; X) := \{(H) \in \Phi(G) : H = G_X \text{ for some } x \in X\}$. Furthermore, for a subgroup, $H \subset G$, put $X^H := \{x \in X : G_X \supset H\}$. As is well known (see, for instance, [26]), the G-action on X induces a natural $W(H)$-action on X^H.

Consider two subgroups, $L \subset H$, of G and put $N(L, H) := \{g \in G : gLg^{-1} \subset H\}$. Clearly, $N(L, H)$ is an $N(H)$-space. Define the number $n(L, H) := |N(L, H)/N(H)|$ having very transparent geometric meaning; it is equal to the cardinality of the set $\{H' : H' \in (H) \text{ and } L \subset H'\}$.

Let X and Y be two G-spaces. A *continuous* map, $f: X \to Y$, is said to be *G-equivariant* (or, simply, a *G-map*) if $f(gx) = gf(x)$ for all $x \in X$ and $g \in G$.

Convention: For a (finite) group, G, we denote by V0, V1, V2, ..., Vr the complete list of all irreducible orthogonal (real) G-representations.

Suppose that V is an orthogonal G-representation (in general, reducible). Then, it is possible to represent V as the following direct sum:

$$V = V_0 \oplus V_1 \oplus \cdots \oplus V_r \tag{5}$$

called the *G-isotypical decomposition* of V, where the *isotypical components* V_k are *modeled* on the irreducible G-representations, $\mathcal{V}_k$. In other words, the component, V_k, is the minimal subrepresentation of V containing all the irreducible subrepresentations of V that are equivalent to $\mathcal{V}_k$. Notice that if $T : V \to V$ is a G-equivariant linear operator, then $T(V_k) \subset V_k$ for all k. Furthermore, denote by $GL^G(V)$ the set of all linear G-equivariant isomorphisms, $T : V \to V$.

Let $S([a,b]; V)$ be a Banach space of reasonable (e.g., continuous, differentiable, Sobolev differentiable, *etc.*) functions, $[a, b] \to V$, where V is an orthogonal G-representation. Then, $S([a,b]; V)$ can be equipped with the structure of a Banach G-representation by letting:

$$(g, u)(t) := g(u(t)), \qquad g \in G, u \in S([a, b]; V) \tag{6}$$

Combining Equations (5) and (6) yields the isotypical decomposition:

$$S([a, b]; V) = S([a, b]; V_0) \oplus S([a, b]; V_1) \oplus ... \oplus S([a, b]; V_r) \tag{7}$$

2.2. Equivariant Degree for Multivalued Vector Fields

In order to treat implicit symmetric BVPs, we will use an extension of the equivariant degree without free parameters to multivalued compact equivariant vector fields with compact convex images. Up to several standard steps, such an extension is very simple (see, for example, [12]). Therefore, below, we will only outline the key steps of the construction (we refer the reader to Appendix 1 of the present paper, where the axiomatic definition for single-valued fields is presented).

Let E be a Banach space. Denote by C (E) (respectively K_C(E)) the family of all non-empty convex subsets of E (respectively all non-empty convex and compact subsets of E). Let X be a subset of a Banach space, Y. A map, $F : X \to C$(E) (respectively $F : X \to K_C$(E)), is called a *multivalued map with convex values from X to* E (respectively *multivalued map with convex and compact values from X to* E).

A multivalued map, $F : X \to K_C$(E), is said to be *upper semi-continuous* (in short, u.s.c.) if for every open set, $U \subset$ E, the set, $\{x \in X : F(x) \subset U\}$, is open in X. A u.s.c multivalued map, $F : X \to K_C$(E), is called *compact* if, for any bounded set, $S \subset X$, the closure of $\bigcup_{x\in S} F(x)$ is compact in E. In what follows, we will write $F \in M_K$ to indicate that F is a u.s.c. compact multivalued map with non-empty compact convex values.

Assume now that E and Y are isometric Banach G-representations, $X \subset$ Y is a G-invariant set and $F \in M_K$ is a multivalued map from X to E. Then, F is called *G-equivariant* if $F(gx) = gF(x)$ for all $x \in X$ and $g \in G$, and we write $F \in \mathscr{M}^G_{\mathscr{X}}$. Observe that if $f : X \to$ E is a single-valued compact G-equivariant map, then it can also be considered as the multivalued map

$Ff(x) := \{f(x)\}$, $x \in X$. Clearly, Ff is u.s.c. (as a multivalued map) and, therefore, $F_f \in \mathcal{M}^G_{\mathcal{K}}$.

Let $\Omega \subset$ E be an open bounded *G*-invariant subset. Similarly to the single-valued case, a multivalued map, $F : \Omega \to K_C(E)$, is called an *Ω-admissible compact G-equivariant field* if the following two conditions are satisfied:

(i) there exists $F : \overline{\Omega} \to K_c(\mathbb{E})$, such that $F \in \mathcal{M}^G_{\mathcal{K}}$ and $\mathfrak{F}(x) = x - F(x)$ for all $x \in \overline{\Omega}$;

(ii) for all $x \in \partial\Omega, 0 \notin x - F(x), i.e., x \notin F(x)$ (by the same token, *F* has no fixed-points in $\partial\Omega$).

In such a case, (F,Ω) is called an *admissible G-pair* in E. Denote by $\mathcal{A}\mathcal{M}^G_{\mathcal{K}}(\mathbb{E})$ the set of all such admissible *G*-pairs in Eand put:

$$\mathcal{A}\mathcal{M}^G_{\mathcal{K}} := \bigcup_{\mathbb{E}} \mathcal{A}\mathcal{M}^G_{\mathcal{K}}(\mathbb{E}) \tag{8}$$

(here, the union is taken over all isometric Banach *G*-representations).

Take $(\mathfrak{F}_0, \Omega), (\mathfrak{F}_1, \Omega) \in \mathcal{A}\mathcal{M}^G_{\mathcal{K}}$. Then, F_0, F_1 are said to be *equivariantly Ω-admissibly homotopic* if there exists a multivalued map, $H : [0,1] \times \Omega \to K_C(E)$, $H \in \mathcal{M}^G_{\mathcal{K}}$, such that:

(a) $\mathfrak{F}_i(x) = x - H(i, x), i = 0, 1,$ for all $x \in \overline{\Omega}$;

(b) $x \notin H(t, x)$ for all $(t, x) \in [0, 1] \times \partial\Omega$.

Lemma 2.1. (*cf.* [12,27]).

(i) For any $(\mathfrak{F},\Omega)\in\mathscr{A}\mathscr{M}^G_{\mathscr{K}}(\mathbb{E})$, there exists an equivariant Ω-admissible homotopy, $H:[0,1]\times\overline{\Omega}\to\mathscr{K}_c(\mathbb{E})$, $H\in\mathscr{M}^G_{\mathscr{K}}$, such that H(0,·)=F and H(1,·)=Ff, where $f:\overline{\Omega}\to\mathbb{E}$ is a single-valued field.

(ii) Let $H:[0,1]\times\overline{\Omega}\to\mathscr{K}_c(\mathbb{E})$, $H\in\mathscr{M}^G_{\mathscr{K}}$ be an equivariant Ω-admissible homotopy, such that H(0,·)=Ff0 and H(1,·)=Ff1, where $f_0,f_1:\overline{\Omega}\to\mathbb{E}$ are (compact) single-valued fields. Then, there exists a single-valued equivariant Ω-admissible homotopy joining f0 and f1.

Lemma 2.1 allows us to extend the G-equivariant degree defined for single-valued admissible G-pairs to the fields from AMGK. Namely, take $(\mathfrak{F},\Omega)\in\mathscr{A}\mathscr{M}^G_{\mathscr{K}}(\mathbb{E})$. Find an admissible G-pair (f,Ω) with single-valued $f:\overline{\Omega}\to\mathbb{E}$, such that Ffis equivariantly Ω-admissibly homotopic to F (cf. Lemma 2.1(i)), and put:

$$G\text{-Deg}(\mathfrak{F},\Omega)\stackrel{\text{def}}{=}G\text{-deg}(f,\Omega) \qquad (9)$$

Using Lemma 2.1(ii), one can easily verify that G-Deg(F,Ω), defined by (9), is independent of a choice of a single-valued representative, *f*. Moreover, by applying the standard argument, one can show that G-Deg satisfies the standard properties. More precisely:

Theorem 2.2. (*cf.* [12]).

There exists a unique map, $\mathscr{A}\mathscr{M}_{\mathscr{X}}^{G} \to A(G)$, which assigns to every admissible G-pair (F,Ω) an element, G-Deg(F,Ω)∈A(G), called the G-equivariant degree (or, simply, G-degree) of F on Ω:

$$G\text{-Deg}(\mathfrak{F},\Omega) = \sum_{(H_i)\in\Phi(G)} n_{H_i}(H_i) = n_{H_1}(H_1) + \cdots + n_{H_m}(H_m) \tag{10}$$

satisfying (among others) the following properties:

(MG1) (Existence) If G-Deg(F, Ω) 6= 0, i.e., there is in Equation (10) a non-zero coefficient, n_{H_i}, then

∃x∈Ω, such that 0 ∈ F(x) and $(G_x) \geq (H_i)$.

(MG2) (Additivity) Let Ω_1 and Ω_2 be two disjoint open G-invariant subsets of Ω, such that, for any $x \in \Omega \setminus (\Omega_1 \cup \Omega_2)$, one has $0 \notin F(x)$. Then:

$$G\text{-Deg}(\mathfrak{F},\Omega) = G\text{-Deg}(\mathfrak{F},\Omega_1) + G\text{-Deg}(\mathfrak{F},\Omega_2)$$

(MG3) **(Homotopy)** If H : [0, 1] × Ω → $K_c(E)$ is an Ω-admissible G-homotopy of multivalued G-equivariant compact fields, then:

$$G\text{-Deg}(\mathfrak{H}_t,\Omega) = \mathit{constant}, \quad t \in [0,1]$$

(MG4) **(Normalization)** *Let* Ω *be a G-invariant open bounded neighborhood of zero in* E. *Then:*

$$G\text{-Deg}(\mathrm{Id}\,,\Omega) = 1 \cdot (G)$$

(MG5) **(Multiplicativity)** *For any* $(\mathfrak{F}_1, \Omega_1), (\mathfrak{F}_2, \Omega_2) \in \mathscr{A}\mathscr{M}^G_{\mathscr{K}}$

$$G\text{-Deg}(\mathfrak{F}_1 \times \mathfrak{F}_2, \Omega_1 \times \Omega_2) = G\text{-Deg}(\mathfrak{F}_1, \Omega_1) \cdot G\text{-deg}(\mathfrak{F}_2, \Omega_2)$$

where the multiplication "·" is taken in the Burnside ring, A(G) (see Appendix 1, Subsection A1.1.).

For the equivariant topology/representation theory background, we refer the reader to [26,28,29,30]. For all the "multivalued" backgrounds frequently used here, we refer the reader to [27,31]. The detailed exposition of the equivariant degree theory can be found in [22,25].

3. SYMMETRIC DIFFERENTIAL INCLUSIONS

3.1. Basic Definitions and Facts

To formulate a result on (symmetric) multivalued BVPs, recall some standard notions and facts.

For any Banach space, E, the set, K_C(E), of all nonempty compact convex sets in E can be equipped with the so-called *Hausdorff metric*, $D(\cdot,\cdot)$. To be more specific, if $A, B \in K_C$(E), put:

$$d(A,B) := \inf\{r > 0 : A \subset B_r(0) + B\}; \quad D(A,B) := \max\{d(A,B), d(B,A)\}$$

(here, $B_r(0)$ stands for the ball of radius *r* centered at the origin). One can easily verify that the function *D* is indeed a metric on K_C(E).

Definition 3.1. Let $\Omega \subset R^n \oplus E$ be an open set. A multivalued map, $F : \overline{\Omega} \to \mathcal{K}_c(\mathbb{E})$, is said to be *measurable* if, for every open set, $U \subset \mathcal{K}_c(\mathbb{E})$ (in the topology induced by the Hausdorff metric), the inverse image:

$$F^{-1}(U) := \{x \in \overline{\Omega} : F(x) \in U\}$$

is Lebesgue measurable.

Definition 3.2. A multivalued map, $F:[0,1]\times R^m \to K_c(R^n)$, is called a *Carathéodory* if it satisfies the following two conditions:

(i) for every $u \in R^m$, the multivalued map $F(\cdot, u)$ is measurable;

(ii) for every $t \in [0, 1]$, the multivalued map $F(t, \cdot)$ is upper semicontinuous.

The following result is well-known (see [32,33]) and plays an important role in our considerations.

Proposition 3.3. *Let* $F : [0,1] \times R^m \to K_c(R^n)$ *be a Carathéodory multivalued map satisfying the following condition:*

- *For any bounded set,* $B \subset \mathbb{R}^m$, *there exists* $\varphi_B \in L^1([0,1];\mathbb{R})$, *such that*:

$$\|F(t,v)\| := \sup\{\|u\| : u \in F(t,v),\ t \in [0,1],\ v \in B\} \leq \varphi_B(t)\}$$

Then, the formula:

$$N_F(v)(t) := \{u \in L^2([0,1],\mathbb{R}^n) : u(t) \in F(t,v(t)) \text{ a.e. } t \in [0,1]\}$$

defines a continuous map from $C([0,1),\mathbb{R}^m)$ *to* $K_c(L^2([0,1],\mathbb{R}^n)$.

3.2. Hypotheses

Put $V:=\mathbb{R}^n$. We are interested in studying the BVP for second order differential inclusion of the type:

$$\begin{cases} \ddot{y} \in Cy(t) + F(t, y(t), \dot{y}(t)) & \text{for a.e. } t \in [0,1] \\ y(0) = 0 = y(1) \end{cases} \tag{11}$$

where $F : [0, 1] \times V \times V \to K_C(V)$ and $C : V \to V$ is a linear operator. As usual, the differentiation is understood in the sense of Sobolev derivatives. We need the following adaptation of the Hartman-Nagumo conditions (*cf.* [2,4]) for the multivalued map, $C + F$.

(H0) F is a Carathéodory map satisfying condition (A), and there exists a constant $R > 0$, such that:

(H1) for any $v_0, w_0 \in V$ satisfying $w_0 \bullet v_0 = 0$, there is $\delta = \delta(v_0, w_0) > 0$, such that

$$\|v_o\| > R \implies \operatorname*{ess\,inf}_{t\in[0,1]} \left\{ v \bullet u + \|w\|^2 \ : \ u \in Cv + F(t, v, w),\ (v, w) \in D_\delta \right\} > 0$$

where $D_\delta := \{(v, w) \in V \times V \ : \ \|v - v_o\| + \|w - w_o\| < \delta\}$;

(H2) there exist α, $\kappa > 0$, such that, for all $v, w \in V$:

$$\|v\| \le R \implies \|Cv + F(t, v, w)\| \le \alpha(v \bullet u + \|w\|^2) + \kappa$$

for a.e. $t \in [0,1]$ and all $u \in Cv + F(t,v,w)$;

(H3) there is a function, $\beta : [0,\infty) \to (0,\infty)$, such that the function, $s \mapsto \frac{s}{\beta(s)},\ s \in [0, \infty)$, belongs to $L^\infty_{loc}[0, \infty)$, $\int_0^\infty \frac{s}{\beta(s)} ds = \infty$, and for all $v, w \in V$ with $\|v\| \le R$,

$$\|Cv + F(t, v, w)\| \le \beta(\|w\|) \quad \text{for a.e. } t \in [0, 1]$$

In addition, we will assume that problem (11) is asymptotically linear at the origin and the linearization at the origin is non-degenerate, *i.e.*:

(H4) $\lim_{(v,w)\to(0,0)} \frac{\|F(t,v,w)\|}{\|(v,w)\|} = 0$ uniformlywithrespecttot∈[0,1];

(H5) the linear system: $\begin{cases} \ddot{y} = Cy \\ y(0) = 0 = y(1) \end{cases}$

has only the trivial solution, $y \equiv 0$, *i.e.*, $\sigma(C) \cap \{-\pi^2 n^2 : n \in \mathbb{N}\} = \emptyset$,
where σ(C) stands for the spectrum of *C*.

Finally, we assume that *V* is a coordinate permutation *G*-representation, *i.e.*, there is a homomorphism, *σ* : G→Sn, such that:

$$g(x_1, x_2, ..., x_n)^T = (x_{\sigma(g)(1)}, x_{\sigma(g)(2)}, ..., x_{\sigma(g)(n)})^T \quad (12)$$

(here, Sn stands for the symmetric group of *n* elements). Moreover, we will always assume that dimVG=1, *i.e.*:

$$V^G = \{(x, x, ..., x)^T : x \in \mathbb{R}\} \quad (13)$$

We make the following assumptions with respect to *F*:

(H6) the multivalued map, F:[0,1]×V×V→Kc(V), is *G*-equivariant, *i.e.*:

$$\forall_{g\in G} \forall_{v,w\in V} \forall_{t\in[0,1]} \quad F(t, gv, gw) = gF(t, v, w)$$

(as usual, G is supposed to act trivially on [0,1]), and the linear map, C:V→V, is G-equivariant, as well.

It follows immediately from condition (H6) that the multivalued map, $F^G : [0,1] \times V^G \times V^G \to \mathscr{K}_c(V^G)$, given by:

$$F^G(t,v,w) := F(t,v,w) \cap V^G, \quad t \in [0,1],\ v,w \in V^G$$

is well-defined and u.s.c. We assume additionally:

(H7) for every non-zero, $v_0 \in V^G$, there is $\delta = \delta(v_0) > 0$, such that:

$$\operatorname*{ess\,inf}_{t\in[0,1]} \left\{ vu + w^2 \ :\ u \in Cv + F^G(t,v,w),\ (v,w) \in D_\delta \right\} > 0$$

Where $D_\delta := \{(v,w) \in V^G \times V^G : |v - v_o| + |w| < \delta\}$

The simple observation, following below, will be essentially used in the sequel.

Lemma 3.4. *Under assumptions (H0)–(H7), the differential inclusion*:

$$\begin{cases} \ddot{x} \in C^G x + F^G(t, x, \dot{x}) & a.e.\ t \in [0, 1] \\ x(0) = 0 = x(1) \end{cases} \tag{14}$$

where CG:=C|VG, *has only the trivial solution,* x≡0.

Proof: Assume for contradiction that x:[0,1]→R=:VG is a solution to (14), such that $r(t) := \frac{1}{2}x^2(t)$ has a positive maximum, *i.e.*, (see the boundary conditions) there is to∈(0,1), such that:

$$r(t_o) = \max\{r(t) : t \in [0, 1]\} > 0$$

Then, $\dot{r}(t_o) = x(t_o)\dot{x}(t_o) = 0$, which implies:

$$\dot{x}(t_o) = 0 \tag{15}$$

Since *x* is a solution to (14), r¨(·)=x(·)x¨(·)+x˙²(·)∈L2 ([0,1]; R) and x¨(t)∈C^Gx(t)+F^G(t,x(t),x˙(t)) for a.e. t∈[0,1]. In particular, *x* is a C¹-smooth function; therefore (see (15)), t→to implies (x(t),x˙(t))→(x(to),0). Hence, there exist α>0 and η>0, such that:

$$\operatorname*{ess\,inf}_{t\in A_\eta}\left\{x(t)u+\dot{x}^2(t)\ :\ u\in Cx(t)+F(t,x(t),\dot{x}(t))\right\}>\alpha>0$$

where Aη:={t∈[0,1]:|to−t|<η} (*cf.* condition (H7)). Thus, for almost every t∈Aη:

$$\ddot{r}(t)=x(t)\ddot{x}(t)+\dot{x}^2(t)>\alpha>0$$

which implies that r˙(t) is increasing for t∈Aη, t>t0 and decreasing for t∈Aη, t<to. However, this is a contradiction with the assumption that r(to) is a maximal value of *r*.

3.3. Operator Reformulation in Functional Spaces and the Existence of Multiple Symmetric Solutions: Abstract Result

Take the Sobolev space, E:=H2([0,1];V), equipped with the norm:

$$\|u\|_{\mathbb{E}}:=\|u\|_{2,2}=\left[\int_0^1\left(u(t)\bullet u(t)+\dot{u}(t)\bullet u(t)+\ddot{u}(t)\bullet\ddot{u}(t)\right)dt\right]^{\frac{1}{2}}$$

(16)

the space: F:=L2([0,1];V)×V2=L2([0,1];V)×V×V equipped with the usual product norm:

$$\|(f,p,q)\|_{\mathbb{F}}:=\max\{\|f\|_{L^2([0,1];V)},\|p\|_V,\|q\|_V\}\quad(17)$$

and the space, C([0,1];V2), of continuous functions from [0,1] to V2 equipped with the norm:

$$\|(u,v)\|_\infty := \max\{\|u\|_\infty, \|v\|_\infty\} \tag{18}$$

where u,v∈C([a0,a1];V). Define the operators:

$$j : \mathbb{E} \to C([0,1];V^2), \quad j(u) = (j_1(u), j_2(u)) := (u, \dot{u}), \quad u \in \mathbb{E} \tag{19}$$

$$L : \mathbb{E} \to \mathbb{F}, \quad Lu := (\ddot{u}, u(a_0), u(a_1)), \quad u \in \mathbb{E} \tag{20}$$

Observe that both jk:E→C([0,1];V), k=1,2 are compact; therefore, *j* is compact, as well.

Furthermore, define the multivalued map $N_F : C([0,1];V^2) \to \mathscr{K}_c(\mathbb{F})$, by

$$N_F(v,w)(t) := \{u \in \mathbb{F} \; : \; u(t) \in F(t, v(t), w(t)) \text{ a.e. } t \in [0,1]\} \times \{(0,0)\}$$

and the operator, C:C([0,1];V2)→F, by:

$$\mathcal{C} : C([0,1];V^2) \to \mathbb{F}, \text{ by:} \tag{21}$$

Since the operator *L* is an isomorphism, the differential inclusion (11) can be reformulated as the following fixed-point problem:

$$u \in L^{-1}(\mathcal{C} + N_F)(j(u)), \quad u \in \mathbb{E} \tag{22}$$

Remark 3.5. (i) The *G*-action on *V* induces in a natural way the *G*-actions on E, C([0,1];V2) and F. For example, the *G*-action on C([0,1];V2) is given by the formula:

$$\forall_{g\in G}\ \ \forall_{u=(u_1,u_2)\in C([0,1];V^2)}\ \ g(u) := (g(u_1(t)), g(u_2(t)))$$

where t∈[0,1].

(ii) The multivalued map, $\mathfrak{F} : \mathbb{E} \to \mathscr{K}_c(\mathbb{E})$, given by

$$\mathfrak{F}(u) := u - L^{-1}(C + N_F)(j(u)), \quad u \in \mathbb{E}$$

is a compact *G*-equivariant multivalued field (*cf.* condition (H6) and compactness of the operator, *j*).

(iii) The map, A:E→E, given by:

$$\mathscr{A}(u) := u - L^{-1} \circ C(j(u)), \quad u \in \mathbb{E} \tag{23}$$

is a *G*-equivariant compact linear field on E, and moreover, by assumption (H5), it is an isomorphism. In particular, (A,Ωδ), where Ωδ:={u∈E:‖u‖2,2<δ} is an admissible *G*-pair for any δ>0; therefore, the *G*-equivariant degree, G-Deg(A,Ωδ)∈A(G), is correctly defined for any δ>0.

(iv) By conditions (H4) and (H5), there exists δo>0 and a *G*-equivariant Ωδo-admissible homotopy joining F and $\mathscr{A}$. Therefore, G-Deg($\mathscr{A}$,Ωδo)=G-Deg(F,Ωδo) (*cf.* property (MG3)).

Put:

$$\omega(C, F) := (G) - G\text{-}\mathrm{Deg}(\mathscr{A}, \Omega_{\delta_o}) \tag{24}$$

Theorem 3.6. *Let V be an orthogonal G-representation satisfying* (12) *and* (13). *Assume F satisfies (H0)–(H7) and let (cf.* (24))) ω(C,F)≠0, *i.e.,*

$$\omega(C,F) = n_1(H_1) + n_2(H_2) + \cdots + n_m(H_m), \quad n_j \neq 0, \; j = 1,2,\ldots,m \tag{25}$$

Then:

- *for every* j=1,2,⋯,m, *there exists a non-zero solution,* u∈E, *to (11), such that* (Gu)≥(Hj).
- *If, in addition,* (Hj) *is a maximal orbit type in* V\VG, *then* (Gu)=(Hj).

Proof: Using conditions (H0)–(H3) and following the standard argument (see, for example, [2,4,23]), one can provide *a priori* estimates for solutions to (22). More precisely, there exists R>0 large enough, such that the multivalued field, F, is*G*-equivariantly admissibly homotopic to Id on the ball ΩR:={u∈E:‖u‖2,2<R}. In particular (see properties (MG3) and (MG4) of the equivariant degree for multivalued fields):

$$G\text{-Deg}(\mathfrak{F}, \Omega_R) = (G) \tag{26}$$

Take δo>0 provided by Remark 3.5(iv). By (MG2), (24) and (26):

$$G\text{-Deg}(\mathfrak{F}, \Omega_R \setminus \Omega_{\delta_o}) = G\text{-Deg}(\mathfrak{F}, \Omega_R) - G\text{-Deg}(\mathfrak{F}, \Omega_{\delta_o}) = \omega(C,F) \tag{27}$$

Combining (25) and (27) with the existence property (MG4) yields Statement (a).

To establish Statement (b), it is enough to combine Statement (a) with assumption (H7) and Lemma 3.4. □

4. SYMMETRIC IMPLICIT BOUNDARY VALUE PROBLEMS

4.1. General Result

In this section, we will apply Theorem 3.6 to study problem (3) in the symmetric setting. Below, we formulate assumptions on *h*. The following condition essentially allows a passage from the "implicit" problem to the single-valued "explicit" one via multivalued equivariant homotopy techniques.

h:[0,1]×V×V×V→V is a Carathéodory function and there exist a Carathéodory function,

α:[0,∞)×V×V→[0,∞), and a constant, 0≤c<1, such that:

$$\|h(t,u,v,w)\| \leq \alpha(t,u,v) + c\|w\| \quad \text{for a.e. } t \in [0,1], \text{ and for all } u,v,w \in V \tag{28}$$

$$\lim_{(u,v)\to(0,0)} \alpha(t,u,v) = 0 \quad \text{uniformly with respect to } t \in [0,1] \tag{29}$$

and:

$$\|h(t,u,v,w_1) - h(t,v,u,w_2)\| \le \|w_1 - w_2\| \quad \text{for a.e. } t \in [0,1]$$

(30)

and for all u,v∈V, w1, w2∈Br:={w∈V:‖w‖<r}, where $r := r(t,u,v) = \frac{\alpha(t,u,v)}{1-c}$

As is very well-known, the set of fixed points of a non-expansive map is *convex*. The following statement was proven in [15]:

Lemma 4.1. *Suppose that a Carathéodory function,* h: [0,1]×V×V×V→V, *satisfies condition (A0). Then, the multivalued map,* F̃:[0,1]×V×V→Kc(V), *given by*:

$$\widetilde{F}(t,u,v) := \{w \in B_{r(t,u,v)} : w = h(t,u,v,w)\} \quad (31)$$

where r(t,u,v) *is given in condition (A0), is well-defined and satisfies the Carathéodory condition.*

Next, three conditions present the adaptation of the Hartman-Nagumo conditions for the implicit BVP. Namely, we assume that there exists R>0, such that:

(A1) for any uo,vo∈V satisfying uo·vo=0, there exists δ=δ(uo,vo)>0, such that if ‖yo‖>R, then:

$$0 < \operatorname*{ess\ inf}_{t \in [0,1]} \{u \bullet h(t, u, v, w) : \|(u, v) - (u_o, v_o)\| \leq \delta \text{ and } \|w\| \leq r\} \tag{32}$$

where r:=r(t,u,v) is given in (A0);

(A2) There exist constants, κ, α>0, such that:

$$\|h(t, u, v, w)\| \leq 2\alpha(u \bullet h(t, u, v, w) + \|v\|^2) + \kappa \quad \text{a.e. } t \in [0, 1] \tag{33}$$

and for all *u*, *v*, w∈V with ‖u‖≤R and ‖w‖≤r:=r(t,u,v) (where r (t,u,v) is given in the condition (A0));

(A3) There is a function, φ:[0,∞)→(0,∞), such that $s \mapsto \frac{s}{\varphi(s)}$, $s \in [0, \infty)$, belongs to $L^{\infty}_{loc}[0, \infty)$ $\int_0^{\infty} \frac{s}{\varphi(s)} ds = \infty$, and: and:

$$\|h(t, u, v, w)\| \leq \varphi(\|v\|) \quad \text{for a.e. } t \in [0, 1] \tag{34}$$

and *u*, *v*, w∈V, with ‖u‖≤R and ‖w‖≤r:=r(t,u,v) (where r(t,u,v) is given in the condition (A0)).

In addition, we will assume that problem (3) is asymptotically linear at the origin and that the linearization at the origin is non-degenerate. More precisely:

(A4) For any $t \in [0, 1]$, the function, $h : [0, 1] \times V \times V \times V \to V$, is differentiable at $(t, 0, 0, 0)$; also, $D_t h(t, 0, 0, 0) \equiv 0 \equiv D_v h(t, 0, 0, 0)$, and $D_u h(t, 0, 0, 0) =: A$, $D_w h(t, 0, 0, 0) =: B$, with det*A* 6= 0 and det(Id −*B*) 6= 0;

(A5) the *characteristic equation*, $\det_{\mathbb{C}} \Delta(\lambda) = 0, \lambda \in \mathbb{C}$, where $\Delta(\lambda) := \lambda[\mathrm{Id}-B]-A$, associated with the system linearized at $(t, 0, 0, 0)$, has no characteristic roots of the form $-\pi^2 n^2$ $(n = 1, 2, 3, \ldots)$.

Finally, as in Subsection 3.2, we assume that V is a coordinate permutation G-representation given by (12), and condition (13) is satisfied. Furthermore, assume that:

(A6) the function, $h : [0,1] \times V \times V \times V \to V$, is G-equivariant, *i.e.*:

$$h(t, gu, gv, gw) = gh(t, u, v, w), \qquad \text{for all } t \in [0,1] \text{ and } u,\ v,\ w \in V$$

(A7) the function $h_o = h|_{[0,1]\times V^G \times V^G \times V^G}$ satisfies the condition: for any uo∈V, there is δ=δ(uo)>0, such that:

$$0 < \operatorname*{ess\ inf}_{t\in[0,1]}\{u \bullet h_o(t,u,v,w) : \|(u,v) - (u_o, 0)\| \le \delta \text{ and } \|w\| \le r\} \tag{35}$$

where r:=r(t,u,v) is given in (A0).

Remark 4.2. A careful analysis of the proof of Lemma 4.1 shows that under the assumption that h is G-equivariant, one can construct $\tilde{F}$ to be G-equivariant, as well.

The Lemma, following below, plays an important role in our considerations.

Lemma 4.3. *Suppose that* h:[0,1]×V×V×V→V *satisfies conditions (A0)–(A7). Let* C:V→V *be given by*:

$$C := (\mathrm{Id} - \mathrm{B})^{-1}\mathrm{A} \tag{36}$$

and let F˜ *be a map provided by Lemma 4.1. Define the multivalued map,* F:[0,1]×V×V→Kc(V), *by*:

$$F(t,u,v) := \widetilde{F}(t,u,v) - Cu, \quad u,\, v \in V \tag{37}$$

Then:

(a) F:[0,1]×V×V→Kc(V) *satisfies conditions (H0)–(H7);*

(b) any solution, u∈E, *to (11) is also a solution to (3).*

Proof: In light of [15] (see also (12) and (13), conditions (A6) and (A7)), we need to check only condition (H4), *i.e.*:

$$\lim_{(u,v)\to(0,0)} \frac{\|F(t,u,v)\|}{\|(u,v)\|} = 0 \tag{38}$$

uniformly with respect to t∈[0,1].

Suppose for contradiction that (38) is not true. Then, there exists ε>0 and a sequence, $\{(t_n, u_n, v_n, w_n)\}_{n=1}^{\infty}$, such that $t_n \to t_o$ and (un,vn)→(0,0), where wn∈F(t,un,vn) and:

$$\|w_n\| \geq \varepsilon \|(u_n, v_n)\| \quad \text{for all } n \in \mathbb{N} \tag{39}$$

Therefore, by the definition of *F*:

$$w_n + Cu_n = g(t_n, u_n, v_n, w_n + Cu_n), \quad n \in \mathbb{N}$$

Therefore:

$$w_n + Cu_n = Au_n + B(w_n + Cu_n) + r(t_n, u_n, v_n; u_n + Cw_n)$$

where r(t,u,v,w)/‖(u,v,w)‖→0 as ‖(u,v,w)‖→0 (uniformly with respect to *t*), which leads to:

$$(\mathbf{Id} - B)w_n = r(t_n, u_n, v_n; u_n + Cw_n) \tag{40}$$

Then, since, by (29), ‖wn‖≤α(tn,un,vn)→0 as n→∞, it follows that:

$$\begin{aligned}\lim_{n\to\infty}\frac{(\mathrm{Id}-B)w_n}{\|(u_n,v_n)\|} &= \lim_{n\to\infty}\frac{r(t_n,u_n,v_n;u_n+Cw_n)}{\|(u_n,v_n)\|}\\ &= \lim_{n\to\infty}\frac{r(t_n,u_n,v_n;u_n+Cw_n)}{\|(u_n,v_n,u_n+Cw_n)\|}\cdot\lim_{n\to\infty}\|u_n+Cw_n\|\\ &= 0\cdot 0=0\end{aligned}$$

and we obtain a contradiction with (39).

Combining Lemma 4.3 with Theorem 3.6, one obtains the following:

Theorem 4.4. *Let* h:[0,1]×V×V×V→V *satisfy conditions (A0)–(A7). Assume* C:V→V *is given by (36) and* F:[0,1]×V×V→Kc(V) *is given by (37). Take* ω(C,F) *defined by (19)–(24) and assume:*

$$\omega(C,F)=n_1(H_1)+n_2(H_2)+\cdots+n_m(H_m),\quad n_j\neq 0,\ j=1,2,\ldots,m$$

Then, for every j=1,2,⋯,m, there exists a non-zero solution u∈H2([0,1];V) to (3), such that Gu⊃Hj. In addition, if (Hj) is a maximal orbit type in V\VG, then Gu=Hj.

4.2. General Formula for ω(C,F)

Theorem 4.4 reduces studying symmetric multiple solutions of (3) to the computation of ω(C,F) (or that is the same (*cf.* Remark 3.5, (23) and (24)) as the computation of the equivariant degree, G-Deg(A,Ωδo)∈A(G)). Below, we give a general formula for it.

Assume *V* admits the isotypical decomposition (5). Then, E:=H2([0,1];V) has the following *G*-isotypical decomposition:

$$\mathbb{E} := \mathbb{E}_0 \oplus \mathbb{E}_1 \oplus \cdots \oplus \mathbb{E}_r \tag{41}$$

where:

$$\mathbb{E}_k := \{u \in \mathbb{E} : \forall_{t\in[a_0,a_1]} \quad u(t) \in V_k\}$$

Since A:E→E is *G*-equivariant, it preserves the *G*-isotypical decomposition of E, *i.e.*:

$$\mathscr{A}(\mathbb{E}_k) = \mathbb{E}_k, \quad k = 0, 1, 2, \ldots, r \tag{42}$$

Since $\mathscr{A}$ is a compact linear field, all points of the spectrum of $\mathscr{A}$ are of finite multiplicity, and one can be the only accumulation point of the spectrum of $\mathscr{A}$. Hence, the negative spectrum, σ−(A), is composed of a finite number of eigenvalues (of finite multiplicity). For each λ∈σ−(A), denote by E(λ) the generalized eigenspace of λ and put (*cf.* (64))

$$m_k(\lambda) := \dim\left(E(\lambda) \cap \mathbb{E}_k\right)/\dim\left(\mathcal{V}_k\right), \quad k = 0, 1, 2, \ldots, r$$

to denote the Vk-multiplicity of the eigenvalue, λ. Consequently, one has (see Theorem 5.5) the following formula:

$$G\text{-Deg}\,(\mathscr{A}, \Omega_{\delta_o}) := G\text{-Deg}\,(D\mathfrak{F}(0), \Omega_{\delta_o}) = \prod_{\lambda\in\sigma_-(\mathscr{A})}\prod_{k=0}^{r}(\deg_{\mathcal{V}_k})^{m_k(\lambda)} \tag{43}$$

Formula (43) requires effective computations of the negative spectrum of $\mathcal{A}$. In the next section, we will show that very often, it is a feasible task.

5. EXAMPLES OF IMPLICIT DN-SYMMETRIC BVPS WITH MULTIPLE SOLUTIONS

In this section, we describe a class of examples illustrating Theorem 4.4. Throughout this section, *V* stands for a Dn-representation given by (72) and (73) and admitting the isotypical decomposition (76).

5.1. A Class of Maps Satisfying (A0)–(A7)

We start with describing a class of functions, h:[0,1]×V×V×V→V, satisfying (A0)–(A7).

Let A:V→V be a Dn-equivariant linear operator and let σ(A):={μj:0≤j≤m} denote the spectrum of *A* (*cf.* Theorem 5.6). Assume that μ0=0, *i.e.*:

$$A|_{V^G} = 0 \tag{44}$$

Let ζ:[0,1]×V→R be a Dn-invariant C1-differentiable function, such that:

$$\zeta(t,u) > 0 \tag{45}$$

For two vectors, *u*, v∈V, define:

$$u \cdot v := (u_1 v_1, u_2 v_2, \dots, u_n v_n)^T, \quad u = (u_1, u_2, \dots, u_n)^T, \; v = (v_1, v_2, \dots, v_n)^T \tag{46}$$

and put:

$$u^{l+1} := u \cdot u^l, \quad \text{for} \quad l \geq 1 \tag{47}$$

Let ψ:[0,1]×V→V be a Dn-equivariant C1-differentiable function, such that:

$$\psi(t,u) = (\psi_1(t,u), \psi_2(t,u), \dots, \psi_n(t,u))^T \text{ with } \psi_i(t,u) \geq 0, \; i = 1, 2, \dots, n, \text{ and } \psi(t,0) \tag{48}$$

Define f:[0,1]×V×V→V by:

$$f(t,u,v) = Au + \left(\|v\|^{\beta} \zeta(t,u)\right) v + u^{2p+1} + u \cdot \psi(t,u) \tag{49}$$

where β∈(0,1) and p∈N.

Let g:V×V→V be a function satisfying the following conditions:

(g1) *g* is Dn-equivariant (in particular, continuous);

(g2)there exist real constants, α>0 and 1>c≥0, such that $\|g(v,w)\| \leq \alpha + c\|w\|$ for all $(v,w) \in V \times V$;

(g3 $\|g(v,w_1) - g(v,w_2)\| \leq \|w_1 - w_2\|$
for all $v, w_1, w_2 \in V$;

(g4) $g(0,0) = 0$;

(g5) $g'_v(0,0) = 0$.

The proof of the statement following below is straightforward.

Proposition 5.1. *Let V be a* Dn-*representation given by (72) and (73), and let* f:[0,1]×V×V→V *be given by (49) (cf. (44)-(48)). Let* g:V×V→V *be a function satisfying (g1)-(g5).*

Then, the function, h:[0,1]×V×V×V→V, *defined by* h(t,u,v,w):= f(t,u,v)+g(v,w) *satisfies conditions (A0)—(A7).*

5.2. Example

One can easily construct a wide class of illustrative examples of implicit BVPs for differential systems symmetric with respect to various classical finite groups (including, in particular, arbitrary dihedral groups Dn, a tetrahedral group A4, an octahedral group S4, an icosahedral group A5, *etc.* (see [25])). However, being motivated by simplicity and the transparency of our exposition, we restrict ourselves to one of the simplest non-abelian symmetry groups, namely D4.

Let V:=R4 be a D4-representation given by (72) and (73). Put y=(y1,y2,y3,y4)T∈V and consider the (autonomous) four-dimensional system of second order ODEs:

$$\begin{cases} \ddot{y}_1 = -2ay_1 + ay_2 + ay_4 + \dot{y}_1 e^{y_1y_2y_3y_4}\|\dot{y}\|^\beta + y_1^3 + y_1y_2^2y_4^2 + k\dot{y}_1\|\dot{y}\|^\beta \sin(\ddot{y}_1)(1+\|\dot{y}\|^2)^{-1} \\ \ddot{y}_2 = -2ay_2 + ay_1 + ay_3 + \dot{y}_2 e^{y_1y_2y_3y_4}\|\dot{y}\|^\beta + y_2^3 + y_2y_1^2y_3^2 + k\dot{y}_2\|\dot{y}\|^\beta \sin(\ddot{y}_2)(1+\|\dot{y}\|^2)^{-1} \\ \ddot{y}_3 = -2ay_3 + ay_2 + ay_4 + \dot{y}_3 e^{y_1y_2y_3y_4}\|\dot{y}\|^\beta + y_3^3 + y_3y_2^2y_4^2 + k\dot{y}_3\|\dot{y}\|^\beta \sin(\ddot{y}_3)(1+\|\dot{y}\|^2)^{-1} \\ \ddot{y}_4 = -2ay_4 + ay_1 + ay_3 + \dot{y}_4 e^{y_1y_2y_3y_4}\|\dot{y}\|^\beta + y_4^3 + y_4y_1^2y_3^2 + k\dot{y}_4\|\dot{y}\|^\beta \sin(\ddot{y}_4)(1+\|\dot{y}\|^2)^{-1} \end{cases}$$

(50)

with the boundary conditions y(0) = 0 = y(1) (here, a>0 and will be specified later (*cf.* conditions (a1) and (a2) below), β∈(0,1) and 0<k≤1).

Define:

$$A := \begin{bmatrix} -2a & a & 0 & a \\ a & -2a & a & 0 \\ 0 & a & -2a & a \\ a & 0 & a & -2a \end{bmatrix} \qquad (51)$$

$$f(y,\dot{y}) := \begin{pmatrix} -2ay_1 + ay_2 + ay_4 + \dot{y}_1 e^{y_1y_2y_3y_4}\|\dot{y}\|^\beta + y_1^3 + y_1y_2^2y_4^2 \\ -2ay_2 + ay_1 + ay_3 + \dot{y}_2 e^{y_1y_2y_3y_4}\|\dot{y}\|^\beta + y_2^3 + y_2y_1^2y_3^2 \\ -2ay_3 + ay_2 + ay_4 + \dot{y}_3 e^{y_1y_2y_3y_4}\|\dot{y}\|^\beta + y_3^3 + y_3y_2^2y_4^2 \\ -2ay_4 + ay_1 + ay_3 + \dot{y}_4 e^{y_1y_2y_3y_4}\|\dot{y}\|^\beta + y_4^3 + y_4y_1^2y_3^2 \end{pmatrix}$$

(52)

$$g(\dot{y},\ddot{y}) := \begin{pmatrix} k\dot{y}_1\|\dot{y}\|^\beta \sin(\ddot{y}_1)(1+\|\dot{y}\|^2)^{-1} \\ k\dot{y}_2|\dot{y}\|^\beta \sin(\ddot{y}_2)(1+\|\dot{y}\|^2)^{-1} \\ k\dot{y}_3\|\dot{y}\|^\beta \sin(\ddot{y}_3)(1+\|\dot{y}\|^2)^{-1} \\ k\dot{y}_4\|\dot{y}\|^\beta \sin(\ddot{y}_4)(1+\|\dot{y}\|^2)^{-1} \end{pmatrix} \qquad (53)$$

Clearly, f (respectively g) is of the form of (49) (respectively it satisfies conditions (g1)–(g5)). Therefore, by Proposition 5.1, h:=f+g satisfies conditions (A0)–(A7). Therefore, in order to apply Theorem 4.4 to find non-zero solutions to (50) and to describe their symmetries, one needs to effectively use formula (43).

Observe that the (symmetric) spectral properties of the linearization A:=DF(0) are completely determined by σ(A)and σ(L) (*cf.* (23)). One has the following D4-isotypical decomposition of *V*:

$$V := \mathcal{V}_0 \oplus \mathcal{V}_1 \oplus \mathcal{V}_3 \tag{54}$$

(see Appendix 2 for the used notations), which implies:

$$\mathbb{E} = \mathbb{E}_0 \oplus \mathbb{E}_1 \oplus \mathbb{E}_3, \quad \mathbb{E}_j := H^2([0,1]; \mathcal{V}_j), \quad j = 0, 1, 3$$

In Figure 1, we show the hierarchy of the orbit types in E.

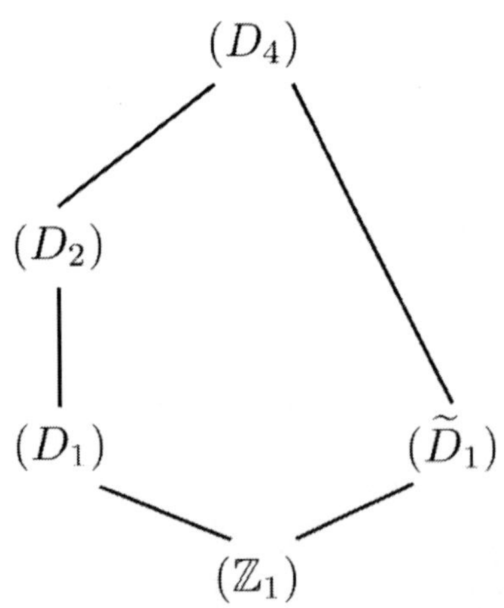

Figure 1. Orbit types in E.

The matrix, *A*, has the eigenvalues, λ0=0, λ1= −2a and λ3=−4a (see Appendix 2 for more details), with the eigenspaces, V0, V1 and V3, respectively. The spectrum of the (D4-equivariant) operator, $\mathscr{A}$, is:

$$\sigma(\mathscr{A}) := \left\{1, 1 - \frac{2a}{\pi^2 k^2}, 1 - \frac{4a}{\pi^2 k^2} : k \in \mathbb{N}\right\} \tag{55}$$

We make the following assumptions regarding σ(A) (take, for example, a=5.5):

(a1) $0 \notin \sigma(\mathscr{A})$;
(a2) $\sigma_-(\mathscr{A}) := \{\mu \in \sigma(\mathscr{A}) : \mu < 0\} = \{1 - \frac{2a}{\pi^2}, 1 - \frac{4a}{\pi^2}\}$ (for simplicity).

Then, formula (43) reads as follows:

$$D_4\text{-Deg}(\mathscr{A}, B_{\delta_o}) = \deg_{\mathcal{V}_1} \cdot \deg_{\mathcal{V}_3} \tag{56}$$

(*cf.* Appendix 2). Combining (56) with (70), (71) and the multiplication table for the Burnside ring, A(D4) (see, for example, Table 1 in [11]), one obtains:

$$\begin{aligned} D_4\text{-Deg}(\mathscr{A}, B_{\delta_o}) &= \deg_{\mathcal{V}_1} \cdot \deg_{\mathcal{V}_3} \\ &= \big((D_4) - (D_1) - (\widetilde{D}_1) + (\mathbb{Z}_1)\big) \cdot \big((D_4) - (D_2)\big) \\ &= (D_4) - (D_2) + (D_1) - (\widetilde{D}_1) \end{aligned}$$

hence (see (24)) $\omega(C, F) = (D_2) - (D_1) + (\widetilde{D}_1)$.. We established the following:

Theorem 5.2. *Given system (50), assume the set (55) satisfies hypotheses* (a1) *and* (a2). *Then, (50) admits at least two (classical) non-zero solutions with symmetry* (D2) *and at least four (classical) solutions with symmetry* (D˜1).

We refer to [10] as an appropriate source of examples of *explicit* Dn-symmetric BVPs that can be converted to *implicit*Dn-symmetric BVPs admitting an arbitrary large number of symmetric solutions.

ACKNOWLEDGEMENTS

Z. Balanov was supported by a grant No. PAPIIT IN-117511-3 from Universidad Nacional Autonoma de Mexico, 01000 Mexico D.F., Mexico.

W. Krawcewicz was supported by Chutian Scholars Program at China Three Gorges University, 8 University avenue, Hubei 443002, China.

CONFLICTS OF INTEREST

The authors declare no conflict of interest.

REFERENCES

1. Coddington, E.; Levenson, N. *Theory of Ordinary Differential Equations*; McGraw-Hill: New York, NY, USA, 1955.

2. Hartman, P. *Ordinary Differential Equations*; John Wiley & Sons: New York, NY, USA, 1964.

3. Mawhin, J. Seventy-Five Years of Global Analysis around the Forced Pendulum Equation. In Proceedings of the Equadiff 9 Proceedings, Conference on Differential Equations and Their

Applications, Brno, Czech Republic, 25–29 August 1997; Agarwal, R.P., Neuman, F., Vosmanský, J., Eds.; Masaryk University: Brno, Czech Republic, 1998; pp. 115–145.

4. Hartman, P. On boundary value problems for systems of ordinary nonlinear second order differential equations. *Trans. Am. Math. Soc.* **1960**, *96*, 493–509.

5. Knobloch, H.W. On the existence of periodic solutions for second order vector differential equations. *J. Differ. Eq.* **1971**,*9*, 67–85.

6. Bebernes, J.W.; Schmidt, K. Periodic boundary value problems for systems of second order differential equations. *J. Differ. Eq.* **1973**, *13*, 33–47.

7. Mawhin, J. Some boundary value problems for Hartman-type perturbations of the ordinary vector *p*-Laplacian. *Nonlinear Anal. TMA* **2000**, *40*, 497–503.

8. Mawhin, J. Periodic Solutions of Systems with *p*-Laplacian-Like Operator. In *Nonlinear Analysis and its Applications to Differential Equations*; Progress in Nonlinear Differential Equations and Their Applications; Birkhäuser Boston: Boston, MA, USA, 2001; Volume 43, pp. 37–63.

9. Mawhin, J.; Urena, A. A Hartman-Nagumo inequality for the vector ordinary *p*-Laplacian and applications to nonlinear boundary value problems. *J. Inequal. Appl.* **2002**, *7*, 701–725.

10. Balanov, Z.; Krawcewicz, W.; Nguyen, M. Multiple Solutions to Implicit Symmetric Boundary Value Problems for Second Order ODEs: Equivariant Degree Approach. *Nonlinear Anal. Theory Methods Appl.* **2013**, *94*, 45–64.

11. Balanov, Z.; Krawcewicz, W.; Rybicki, S.; Steinlein, H. A short treatise on the equivariant degree theory and its applications. *J. Fixed Point Theory Appl.* **2010**, *8*, 1–74.

12. Dzedzej, Z. Equivariant degree of convex-valued maps applied to set-valued BVP. *Cent. Eur. J. Math.* **2012**, *10*, 2173–2186.

13. Petryshyn, W.V. Solvability of various boundary value problems for the equation $x'' = f(t, x, x', x'') - y$. *Pac. J. Math.* **1986**, *122*, 169–195.

14. Erbe, L.; Krawcewicz, W. Nonlinear boundary value problems for differential inclusions $y'' \in F(t, y, y')$. *Ann. Pol. Math.* **1991**, *3*, 195–226.

15. Erbe, L.; Krawcewicz, W.; Kaczynski, T. Solvability of two-point boundary value problems for system of nonlinear differential equations of the form $y'' = g(t, y, y', y'')$. *Rocky Mount. J. Math.* **1990**, *20*, 899–907. [

16. Golubitsky, M.; Stewart, I.N.; Schaeffer, D.G. *Singularities and Groups in Bifurcation Theory*; Applied Mathematical Sciences 69; Springer: New York, NY, USA, 1988; Volume II.

17. Balanov, Z.; Schwartzman, E. Morse complex, even functionals and asymptotically linear differential equations with resonance at infinity. *Topol. Methods Nonlinear Anal.* **1998**, *12*, 323–366.

18. Bartsch, T. Topological Methods for Variational Problems with Symmetries. In *Lecture Notes in Mathematics 1560*; Springer: Berlin, Germany, 1993.

19. Mawhin, J.; Willem, M. *Critical Point Theory and Hamiltonian Systems*; Applied Mathematical Sciences 74; Springer: New York, NY, USA, 1989.

20. Rabinowitz, P.H. *Minimax Methods in Critical Point Theory with Applications to Differential Equations*; American Mathematical Society: Providence, RI, USA, 1986.

21. Dihedral Calculator Home Page. Available online: http://dihedral.muchlearning.org/ (accessed on 31 October 2013).

22. Ize, J.; Vignoli, A. Equivariant Degree Theory. In *De Gruyter Series in Nonlinear Analysis and Applications*; Walter de Gruyter: Berlin, Germany, 2003; Volume 8.

23. Krawcewicz, W.; Wu, J. Theory of Degrees with Applications to Bifurcations and Differential Equations. In *Canadian Mathematical Society Series of Monographs and Advanced Texts*; John Wiley & Sons: New York, NY, USA, 1997.

24. Kushkuley, A.; Balanov, Z. Geometric Methods in Degree Theory for Equivariant Maps. In *Lecture Notes in Mathematics 1632*; Springer-Verlag: Berlin, Germany, 1996.

25. Balanov, Z.; Krawcewicz, W.; Steinlein, H. Applied Equivariant Degree. In *AIMS Series on Differential Equations & Dynamical Systems*; AIMS: Springfield, IL, USA, 2006; Volume 1.

26. Dieck, T. *Transformation Groups*; Walter de Gruyter: Berlin, Germany, 1987.

27. Borisovich, G.; Gelman, B.D.; Myshkis, A.D.; Obukhovskii, V.V. *Introduction to the Theory of Multivalued Mappings and Differential Inclusions*; URSS: Moscow, Russian, 2005.

28. Bredon, G.E. *Introduction to Compact Transformation Groups*; Academic Press: New York, NY, USA, 1972. [

29. Kawakubo, K. *The Theory of Transformation Groups*; The Clarendon Press: Oxford, UK, 1991.

30. Bröcker, T.; Dieck, T. *Representations of Compact Lie Groups*; Springer-Verlag: Berlin, Germany, 1985.

31. Gorniewicz, L. *Topological Fixed Point Theory of Multivalued Mappings*; Springer: Berlin, Germany, 1999; Volume 495.

32. Pruszko, T. Some applications of the topological degree theory to multivalued boundary value problems. *Diss. Math.* **1984**, *229*, 1–48.

33. Pruszko, T. Topological degree methods in multivalued boundary value problems. *Nonlinear Anal.* **1978**, *2*, 263–309.

CHAPTER 5

Half-Step Continuous Block Method for the Solutions of Modeled Problems of Ordinary Differential Equations

A. A. James[1], A. O. Adesanya[2], J. Sunday[3], D. G. Yakubu[4]

[1]Department of Mathematics, American University of Nigeria, Yola, Nigeria

[2]Department of Mathematics, Modibbo University of Technology, Yola, Nigeria

[3]Department of Mathematical Sciences, Adamawa State University, Mubi, Nigeria

[4]Department of Mathematics, Tafawa Balewa Federal University of Bauchi, Bauch State, Nigeria

ABSTRACT

In this paper, we developed a new continuous block method using the approach of collocation of the differential system and interpolation of the power series approximate solution. A constant step length within a half step interval of integration was adopted. We evaluated at grid and off grid points to get a continuous linear multistep method. The continuous linear multistep method is solved for the independent solution to yield a continuous block method which is evaluated at selected points to yield a discrete block method. The basic properties of the block method were investigated and found to be consistent and zero stable hence convergent. The new method was tested on real life problems namely: SIR model, Growth model and Mixture Model. The results were found to compete favorably with the existing methods in terms of accuracy and error bound.

KEYWORDS

Approximate Solution; Interpolation; Collocation; Half Step; Converges; Block Method

1. INTRODUCTION

We consider the numerical solution of first order initial value problems of the form:

$$y' = f(x, y), y(x_0) = y_0 \tag{1}$$

where f is continuous and satisfies Lipchitz's condition that guarantees the uniqueness and existence of a solution.

Problem in the form (1) has wide application in physical science, engineering, economics, etc. Very often, these problems do not have an analytical solution, and this has necessitated the deviation of numerical schemes to approximate their solutions.

In the past, scholars have developed a continuous linear multistep in solving (1). These authors proposed methods with different basis functions and among them were [1-6] to mention a few.

These authors proposed methods ranging from predictor corrector method to discrete block method.

Scholars later proposed block method. This block method has the properties of Runge-kutta method for being self-starting and does not require development of separate predictors or starting values. Among these authors are [7-12]. Block method was found to be cost effective and gave better approximation.

This paper is divided into sections as follows: Section 1 is the introduction and background of the study; Section 2 contains the discussion about the methodology involved in deriving the continuous multistep method and the continuous block method. Section 3 considers the analysis of the block method in terms of the order, zero stability and the region of absolute stability. Section 4 focuses on the application of the new method on some numeric examples and Section 5 is on the discussion of result. We tested our method on first order ordinary differential equations and compared our result with existing methods.

2. METHODOLOGY

Consider power series approximate solution in the form

$$y(x) = \sum_{j=0}^{s+r-1} a_j x^j \qquad (2)$$

where S and r are the number of interpolation and collocation points respectively.

The first derivative of (2) gives

$$y'(x) = \sum_{j=0}^{s+r-1} ja_j x^{j-1} \tag{3}$$

Substituting (3) into (2) gives

$$f(x, y) = \sum_{j=0}^{s+r-1} ja_j x^{j-1} \tag{4}$$

Collocating (3) at $x_{n+s}, s = 0\left(\frac{1}{12}\right)\frac{1}{2}$ and interpolating

(2) at x_n gives a system of non-linear equation in the form

$$AX = U \tag{5}$$

where

$$\begin{bmatrix} 1 & x_n & x_n^2 & x_n^3 & x_n^4 & x_n^5 & x_n^6 & x_n^7 \\ 0 & 1 & 2x_n & 3x_n^2 & 4x_n^3 & 5x_n^4 & 6x_n^5 & 7x_n^6 \\ 0 & 1 & 2x_{n+\frac{1}{12}} & 3x^2_{n+\frac{1}{12}} & 4x^3_{n+\frac{1}{12}} & 5x^4_{n+\frac{1}{12}} & 6x^5_{n+\frac{1}{12}} & 7x^6_{n+\frac{1}{12}} \\ 0 & 1 & 2x_{n+\frac{1}{6}} & 3x^2_{n+\frac{1}{6}} & 4x^3_{n+\frac{1}{6}} & 5x^4_{n+\frac{1}{6}} & 6x^5_{n+\frac{1}{6}} & 7x^6_{n+\frac{1}{6}} \\ 0 & 1 & 2x_{n+\frac{1}{4}} & 3x^2_{n+\frac{1}{4}} & 4x^3_{n+\frac{1}{4}} & 5x^4_{n+\frac{1}{4}} & 6x^5_{n+\frac{1}{4}} & 7x^6_{n+\frac{1}{4}} \\ 0 & 1 & 2x_{n+\frac{1}{3}} & 3x^2_{n+\frac{1}{3}} & 4x^3_{n+\frac{1}{3}} & 5x^4_{n+\frac{1}{3}} & 6x^5_{n+\frac{1}{3}} & 7x^6_{n+\frac{1}{3}} \\ 0 & 1 & 2x_{n+\frac{5}{12}} & 3x^2_{n+\frac{5}{12}} & 4x^3_{n+\frac{5}{12}} & 5x^4_{n+\frac{5}{12}} & 6x^5_{n+\frac{5}{12}} & 7x^6_{n+\frac{5}{12}} \\ 0 & 1 & 2x_{n+\frac{1}{2}} & 3x^2_{n+\frac{1}{2}} & 4x^3_{n+\frac{1}{2}} & 5x^4_{n+\frac{1}{2}} & 6x^5_{n+\frac{1}{2}} & 7x^6_{n+\frac{1}{2}} \end{bmatrix},$$

$$A = \begin{bmatrix} a_0 \\ a_1 \\ a_2 \\ a_3 \\ a_4 \\ a_5 \\ a_6 \\ a_7 \end{bmatrix}, U = \begin{bmatrix} y_n \\ f_n \\ f_{n+\frac{1}{12}} \\ f_{n+\frac{1}{6}} \\ f_{n+\frac{1}{4}} \\ f_{n+\frac{1}{3}} \\ f_{n+\frac{5}{12}} \\ f_{n+\frac{1}{2}} \end{bmatrix}$$

Solving (5) for the a_js and substituting back into (4) gives a continuous multistep method in the form

$$y(x) = \alpha_0 y_n + h\sum_{j=0}^{\frac{1}{2}} \beta_j(x) f_{n+j}, j = 0\left(\frac{1}{12}\right)\frac{1}{2} \tag{6}$$

where $a_0 = 1$ and the coefficients of f_{n+j} gives

$$\beta_0 = \left(\frac{1}{120}\right)\left(124416t^7 - 254016t^6 + 211680t^5 - 92610t^4 + 22736t^3 - 3087t^2 + 210t\right)$$

$$\beta_{\frac{1}{12}} = \left(-\frac{1}{35}\right)\left(-124416t^7 - 241920t^6 - 187488t^5 + 73080t^4 - 14616t^3 + 1260t^2\right)$$

$$\beta_{\frac{1}{6}} = \left(\frac{1}{35}\right)\left(311040t^7 - 57456t^6 + 414288t^5 - 145215t^4 + 24570t^3 - 1575t^2\right)$$

$$\beta_{\frac{1}{4}} = \left(-\frac{1}{105}\right)\left(-1244160t^7 - 2177280t^6 - 1463616t^5 + 468720t^4 - 71120t^3 + 4200t^2\right)$$

$$\beta_{\frac{1}{3}} = \left(\frac{1}{70}\right)\left(622080t^7 - 1028160t^6 + 647136t^5 - 193410t^4 + 27720t^3 - 1575t^2\right)$$

$$\beta_{\frac{5}{12}} = \left(-\frac{1}{35}\right)\left(-124416t^7 + 193536t^6 - 114912t^5 + 32760t^4 - 4536t^3 + 252t^2\right)$$

$$\beta_{\frac{1}{2}} = \left(\frac{1}{105}\right)\left(62208t^7 - 90720t^6 + 51408t^5 - 14175t^4 + 1918t^3 - 105t^2\right)$$

where $t = \frac{x - x_n}{h}$. Solving (6) for the independent solution gives a continuous block method in the form

$$y_{n+k} = \sum_{j=0}^{\mu-1} \frac{(jh)^m}{m!} y_n^{(m)} + h^{\mu} \sum_{j=0}^{s} \sigma_j(x) f_{n+j} \qquad (7)$$

where μ is the order of the differential equation s is the collocation points. Hence the coefficient of f_{n+j} in (7)

$$\sigma_0 = \left(\frac{1}{120}\right)\left(124416t^7 - 254016t^6 + 211680t^5\right.$$
$$\left. -92610t^4 + 22736t^3 - 3087t^2 + 210t\right)$$

$$\sigma_{\frac{1}{12}} = \left(-\frac{1}{35}\right)\left(-124416t^7 - 241920t^6 - 187488t^5\right.$$
$$\left. +73080t^4 - 14616t^3 + 1260t^2\right)$$

$$\sigma_{\frac{1}{6}} = \left(\frac{1}{35}\right)\left(311040t^7 - 57456t^6 + 414288t^5\right.$$
$$\left. -145215t^4 + 24570t^3 - 1575t^2\right)$$

$$\sigma_{\frac{1}{4}} = \left(-\frac{1}{105}\right)\left(-1244160t^7 - 2177280t^6 - 1463616t^5\right.$$
$$\left. +468720t^4 - 71120t^3 + 4200t^2\right)$$

$$\sigma_{\frac{1}{3}} = \left(\frac{1}{70}\right)\left(622080t^7 - 1028160t^6 + 647136t^5\right.$$
$$\left. -193410t^4 + 27720t^3 - 1575t^2\right)$$

$$\sigma_{\frac{5}{12}} = \left(-\frac{1}{35}\right)\left(-124416t^7 + 193536t^6 - 114912t^5\right.$$
$$\left. +32760t^4 - 4536t^3 + 252t^2\right)$$

$$\sigma_{\frac{1}{2}} = \left(\frac{1}{105}\right)\left(62208t^7 - 90720t^6 + 51408t^5\right.$$
$$\left. -14175t^4 + 1918t^3 - 105t^2\right)$$

where $t = \frac{x - x_n}{h}$.

Evaluating (2.5) at $t = \frac{1}{12}\left(\frac{1}{12}\right)\frac{1}{2}$ gives a discrete block formula of the form

$$Y_m = ey_n + hdf\left(y_n\right) + hdf\left(Y_m\right) \tag{8}$$

where $e, d,$ are $r \times r$ matrix where

$$d = \begin{bmatrix} \frac{19087}{725760} & \frac{1139}{45360} & \frac{137}{5376} & \frac{143}{5670} & \frac{3715}{145152} & \frac{41}{1680} \end{bmatrix}^T$$

$$Y_m = \left[y_{n+\frac{1}{12}}, y_{n+\frac{1}{6}}, y_{n+\frac{1}{4}}, y_{n+\frac{1}{3}}, y_{n+\frac{5}{12}}, y_{n+\frac{1}{2}} \right]^T$$

$$e = \begin{bmatrix} 0 & 0 & 0 & 0 & 0 & 1 \\ 0 & 0 & 0 & 0 & 0 & 1 \\ 0 & 0 & 0 & 0 & 0 & 1 \\ 0 & 0 & 0 & 0 & 0 & 1 \\ 0 & 0 & 0 & 0 & 0 & 1 \\ 0 & 0 & 0 & 0 & 0 & 1 \end{bmatrix}$$

$$b = \begin{bmatrix} \frac{2713}{30240} & \frac{-15487}{241920} & \frac{293}{5670} & \frac{-6737}{241920} & \frac{263}{30240} & \frac{-863}{725760} \\ \frac{47}{378} & \frac{11}{15120} & \frac{83}{2835} & \frac{-269}{15120} & \frac{11}{1890} & \frac{-37}{45360} \\ \frac{27}{224} & \frac{387}{8960} & \frac{17}{210} & \frac{-243}{8960} & \frac{9}{1120} & \frac{-29}{26880} \\ \frac{116}{945} & \frac{32}{945} & \frac{376}{2835} & \frac{29}{1890} & \frac{4}{945} & \frac{-2}{2835} \\ \frac{725}{6048} & \frac{2125}{48384} & \frac{125}{1134} & \frac{3875}{48384} & \frac{235}{6048} & \frac{-275}{145152} \\ \frac{9}{70} & \frac{9}{560} & \frac{17}{105} & \frac{9}{560} & \frac{9}{70} & \frac{41}{1680} \end{bmatrix}$$

3. ANALYSIS OF THE BASIC PROPERTIES OF OUR NEW METHOD

3.1. Order of the Method

Let the linear operator $L\{y(x):h\}$ associated with the block formular be defined as

$$L\{y(x):h\} = A^{(0)}Y_m - ey_n - h^{\mu}df(y_n) - h^{\mu}bF(Y_m) \tag{9}$$

expanding in Taylor series and comparing the coefficient of h gives

$$\begin{aligned} L\{y(x);h\} &= c_0 y(x) + c_1 hy'^{(x)} + c_1 hy''^{(x)} \cdots c_p h^p y^p(x) \\ &+ c_{p+1}h^{p+1}y^{p+1}(x) + c_{p+2}h^{p+2}y^{p+2} \end{aligned} \tag{10}$$

Definition:-The linear operator L and the associated continuous linear multistep method (3.1) are said to be of order p if $c_0 = c_1 = c_2 = \cdots = c_p = 0$ and $c_{p+1} \neq 0$ is called the error constant and implies that the local truncation error is given by $t_{n+k} = C_{p+2}h^{p+1}y^{p+1}(x_n) + 0(h^{p+2})$.

For our method;

$$L\{y(x);h\}=\begin{bmatrix} y_{n+\frac{1}{12}} \\ y_{n+\frac{1}{6}} \\ y_{n+\frac{1}{4}} \\ y_{n+\frac{1}{3}} \\ y_{n+\frac{5}{12}} \\ y_{n+\frac{1}{2}} \end{bmatrix}-\begin{bmatrix} 1 & \frac{1}{12}h \\ 1 & \frac{1}{6}h \\ 1 & \frac{1}{4}h \\ 1 & \frac{1}{3}h \\ 1 & \frac{5}{12}h \\ 1 & \frac{1}{2}h \end{bmatrix}[y_n]-\begin{bmatrix} f_n \\ f_{n+\frac{1}{12}} \\ f_{n+\frac{1}{6}} \\ f_{n+\frac{1}{4}} \\ f_{n+\frac{1}{3}} \\ f_{n+\frac{5}{12}} \\ f_{n+\frac{1}{2}} \end{bmatrix}=0$$

Expanding in Taylor series expansion gives

$$\sum_{j=0}^{\infty}\frac{\left(\frac{1}{12}\right)^j}{j!}y_n^j-y_n-\frac{19087}{725760}hy_n'-\sum_{j=0}^{\infty}\frac{h^{j+1}}{j!}y_n^{j+1}\left\{\frac{2713}{30240}\left(\frac{1}{12}\right)^j+\frac{15487}{241920}\left(\frac{1}{6}\right)^j\right.$$
$$\left.-\frac{293}{5670}\left(\frac{1}{4}\right)^j+\frac{6737}{241920}\left(\frac{1}{3}\right)^j+\frac{263}{30240}\left(\frac{5}{6}\right)^j-\frac{863}{725760}\left(\frac{1}{2}\right)^j\right\}$$

$$\sum_{j=0}^{\infty}\frac{\left(\frac{1}{6}\right)^j}{j!}y_n^j-y_n-\frac{1139}{45360}hy_n'-\sum_{j=0}^{\infty}\frac{h^{j+1}}{j!}y_n^{j+1}\left\{\frac{47}{378}\left(\frac{1}{12}\right)^j+\frac{11}{15120}\left(\frac{1}{6}\right)^j\right.$$
$$\left.+\frac{83}{2835}\left(\frac{1}{4}\right)^j-\frac{269}{15120}\left(\frac{1}{3}\right)^j-\frac{11}{1890}\left(\frac{5}{12}\right)^j-\frac{37}{45360}\left(\frac{1}{2}\right)^j\right\}$$

$$\sum_{j=0}^{\infty}\frac{\left(\frac{1}{4}\right)^j}{j!}y_n^j-y_n-\frac{137}{5376}hy_n'-\sum_{j=0}^{\infty}\frac{h^{j+1}}{j!}y_n^{j+1}\left\{\frac{27}{224}\left(\frac{1}{12}\right)^j+\frac{387}{8960}\left(\frac{1}{6}\right)^j\right.$$
$$\left.+\frac{17}{210}\left(\frac{1}{4}\right)^j-\frac{243}{8960}\left(\frac{1}{3}\right)^j+\frac{9}{1120}\left(\frac{5}{12}\right)^j-\frac{29}{26880}\left(\frac{1}{2}\right)^j\right\}$$

$$\sum_{j=0}^{\infty}\frac{\left(\frac{1}{3}\right)^j}{j!}y_n^j-y_n-\frac{143}{5670}hy_n'-\sum_{j=0}^{\infty}\frac{h^{j+1}}{j!}y_n^{j+1}\left\{\frac{116}{945}\left(\frac{1}{12}\right)^j+\frac{32}{945}\left(\frac{1}{6}\right)^j\right.$$

$$\left.+\frac{376}{2835}\left(\frac{1}{4}\right)^j+\frac{29}{1890}\left(\frac{1}{3}\right)^j+\frac{4}{945}\left(\frac{5}{12}\right)^j-\frac{2}{2835}\left(\frac{1}{2}\right)^j\right\}$$

$$\sum_{j=0}^{\infty}\frac{\left(\frac{5}{12}\right)^j}{j!}y_n^j-y_n-\frac{3715}{145152}hy_n'-\sum_{j=0}^{\infty}\frac{h^{j+1}}{j!}y_n^{j+1}\left\{\frac{725}{6048}\left(\frac{1}{12}\right)^j+\frac{2125}{48384}\left(\frac{1}{6}\right)^j\right.$$

$$\left.+\frac{125}{1134}\left(\frac{1}{4}\right)^j+\frac{3875}{48384}\left(\frac{1}{3}\right)^j+\frac{235}{6048}\left(\frac{5}{12}\right)^j-\frac{275}{145152}\left(\frac{1}{2}\right)^j\right\}$$

$$\sum_{j=0}^{\infty}\frac{\left(\frac{1}{2}\right)^j}{j!}y_n^j-y_n-\frac{41}{1680}hy_n'-\sum_{j=0}^{\infty}\frac{h^{j+1}}{j!}y_n^{j+1}\left\{\frac{9}{70}\left(\frac{1}{12}\right)^j+\frac{9}{560}\left(\frac{1}{6}\right)^j\right.$$

$$\left.+\frac{17}{105}\left(\frac{1}{4}\right)^j+\frac{9}{560}\left(\frac{1}{3}\right)^j+\frac{9}{70}\left(\frac{5}{12}\right)^j-\frac{41}{1680}\left(\frac{1}{2}\right)^j\right\}$$

Equating coefficients of the Taylor series expansion to zero yield

$$c_0=c_1=\cdots=c_6=0$$

Hence we arrived at a uniform order 6 for our method with error constants

$$c_7=\left[3.78(-09)2.53(-09)3.49(-09)1.85(-09)1.68(-08)5.56(-08)\right]$$

3.2. Zero Stability

Definition: The block (8) is said to be zero stable, if the roots Zs, $s = 1, 2, \cdots, N$ of the characteristic polynomial $\rho(z)$ defined by $\rho(z)\det\left(zA^{(0)} - E\right)$ satisfies $|z_s| \le 1$ and every root satisfying $|z_s| \le 1$ have multiplicity not exceeding the order of the differential equation. Moreover as $h \to 0, \rho(z) = z^{r-\mu}(z-1)^{\mu}$ where μ is the order of the differential equation, r is the order of the matrix $A^{(0)}$ and E.

For our method

$$\rho(z) = \left[z \begin{bmatrix} 1 & 0 & 0 & 0 & 0 & 0 \\ 0 & 1 & 0 & 0 & 0 & 0 \\ 0 & 0 & 1 & 0 & 0 & 0 \\ 0 & 0 & 0 & 1 & 0 & 0 \\ 0 & 0 & 0 & 0 & 1 & 0 \\ 0 & 0 & 0 & 0 & 0 & 1 \end{bmatrix} - \begin{bmatrix} 0 & 0 & 0 & 0 & 0 & 1 \\ 0 & 0 & 0 & 0 & 0 & 1 \\ 0 & 0 & 0 & 0 & 0 & 1 \\ 0 & 0 & 0 & 0 & 0 & 1 \\ 0 & 0 & 0 & 0 & 0 & 1 \\ 0 & 0 & 0 & 0 & 0 & 1 \end{bmatrix} \right] = 0$$

$\rho(z) = z^5(z-1)$. Hence our method is zero stable.

3.3 Region of Absolute Stability

The block formulated as a general linear method where it is partition in the form

$$\begin{bmatrix} Y \\ Y_{n-i} \end{bmatrix} = \begin{bmatrix} A_1 & B_1 \\ A_2 & B_2 \end{bmatrix} \begin{bmatrix} hf(y) \\ y_n \end{bmatrix}$$

The elements of A_1 and A_2 are obtained from the coefficients of the collocation points, B_1 and B_2 are obtained from the interpolation points.

Applying the test equation $y' = \lambda y$ leads to the recurrence equation

$$y^{i+1} = M(Z)y^i,\ Z = \lambda h,\ i = 1,2,\cdots,\mu-1$$

The stability function is given by

$$M(Z) = B_2 + ZA_2(I - ZA_1)^{-B_1}$$

and the stability polynomial of the method is given as

$$\rho(\lambda, Z) = \det(\lambda I - M(Z))$$

The region of absolute stability of the method is defined as $\rho(\lambda, Z) = 1,\ |\lambda| \le 1.$

For our method, writing the block in partition form gives

$$\left[y_n, y_{n+\frac{1}{12}}, y_{n+\frac{1}{6}}, y_{n+\frac{1}{4}}, y_{n+\frac{1}{3}}, y_{n+\frac{5}{12}}, y_{n+1}\right]^T$$

$$= \begin{bmatrix} 0 & 0 & 0 & 0 & 0 & 0 & 0 & \cdots & 0 & 1 \\ \frac{19087}{725760} & \frac{2713}{30240} & \frac{-15487}{241920} & \frac{293}{5670} & \frac{-6737}{241920} & \frac{263}{30240} & \frac{-863}{725760} & \cdots & 0 & 1 \\ \frac{1139}{45360} & \frac{47}{378} & \frac{11}{15120} & \frac{83}{2835} & \frac{-269}{15120} & \frac{11}{1890} & \frac{-37}{45360} & \cdots & 0 & 1 \\ \frac{137}{5376} & \frac{27}{224} & \frac{387}{8960} & \frac{17}{210} & \frac{-243}{8960} & \frac{9}{1120} & \frac{-29}{26880} & \cdots & 0 & 1 \\ \frac{143}{5670} & \frac{116}{945} & \frac{32}{945} & \frac{376}{2835} & \frac{29}{1890} & \frac{4}{945} & \frac{-2}{2835} & \cdots & 0 & 1 \\ \frac{3715}{145152} & \frac{725}{6048} & \frac{2125}{48384} & \frac{125}{1134} & \frac{3875}{48384} & \frac{235}{6048} & \frac{-275}{145152} & \cdots & 0 & 1 \\ \frac{41}{1680} & \frac{9}{70} & \frac{9}{560} & \frac{17}{105} & \frac{9}{560} & \frac{9}{70} & \frac{41}{1680} & \cdots & 0 & 1 \\ \vdots & \vdots & \vdots & \vdots & \vdots & \vdots & \vdots & \ddots & \vdots & \vdots \\ \frac{19087}{725760} & \frac{2713}{30240} & \frac{-15487}{241920} & \frac{293}{5670} & \frac{-6737}{241920} & \frac{263}{30240} & \frac{-863}{725760} & \cdots & 0 & 1 \\ \frac{41}{1680} & \frac{9}{70} & \frac{9}{560} & \frac{17}{105} & \frac{9}{560} & \frac{9}{70} & \frac{41}{1680} & \cdots & 1 & 1 \end{bmatrix} \begin{bmatrix} hf_n \\ hf_{n+\frac{1}{12}} \\ hf_{n+\frac{1}{6}} \\ hf_{n+\frac{1}{4}} \\ hf_{n+\frac{1}{3}} \\ hf_{n+\frac{5}{12}} \\ hf_{n+\frac{1}{2}} \\ \vdots \\ hf_{n+\frac{1}{12}} \\ hf_{n+\frac{1}{2}} \end{bmatrix}$$

4. REAL LIFE PROBLEMS

4.1. Problem 1: (SIR MODEL)

The SIR model is an epidemiological model that computes the theoretical numbers of people infected with a contagious illness in a closed population over time. The name of this class of models derives from the fact that they involves coupled equations relating the number of susceptible people S(t), number of people infected I(t) and the number of people who have recovered R(t). This is a good and simple model for many infectious diseases including measles, mumps and rubella [13-15]. The SIR model is described by the three coupled equations.

$$\frac{\mathrm{d}s}{\mathrm{d}t} = \mu(1-S) - \beta IS \tag{11}$$

$$\frac{\mathrm{d}I}{\mathrm{d}t} = -\mu I - \gamma I + \beta IS \tag{12}$$

$$\frac{\mathrm{d}s}{\mathrm{d}t} = -\mu R + \gamma I \tag{13}$$

where μ, γ and β are positive parameters. Define y to be

$$y = S + I + R \tag{14}$$

Adding Equations (11)-(13), we obtain the following evolution equations for y

$$y' = \mu(1-y) \tag{15}$$

Taking $\mu = 0.5$ and attaching an initial condition $y(0) = 0.5$ (for a particular closed population), we obtain,

$$y'(t) = 0.5(1-y), y(0) = 0.5 \tag{16}$$

whose exact solution is,

$$y(t) = 1 - 0.5\mathrm{e}^{-0.5t} \tag{17}$$

Applying our new half step numerical scheme (8) to solve SIR model simplified as (17) gives results as shown in **Table 1**.

4.2. Problem 2 (Growth Model)

Let us consider the differential equation of the form;

$$\frac{dN}{dt} = \alpha N,\ N(0) = 1000,\ t \in [0,1] \tag{18}$$

Equation (18) represents the rate of growth of bacteria in a colony. We shall assume that the model grows continuously without restriction. One may ask; how many bacteria are in the colony after some minutes if an individual produces an offspring at an average growth rate of 0.2? We also assume that $N(t)$ is the population size at time t (**Table 2**).

The theoretical solution of (18) is given by;

s(19)

Note that, growth rate $\alpha = 0.2$ in (18).

Applying our new half step numerical scheme (8) to solve the Growth model (17) gives results as shown in **Table 2** [16].

Table 1. Showing results for SIR model problem.

X	Exact Result	Computed Solution	Error in half step method	Error in Sunday *et al.*
0.1	0.5243852877496430	0.5243852877496429	1.110223e−016	5.574430e−012
0.2	0.5475812909820202	0.5475812909820202	0.000000e+000	3.946177e−012
0.3	0.5696460117874711	0.5696460117874710	1.110223e−016	8.183232e−012
0.4	0.5906346234610092	0.5906346234610089	2.220446e−016	3.436118e−011
0.5	0.6105996084642976	0.6105996084642973	3.330669e−016	1.929743e−010
0.6	0.6295908896591411	0.6295908896591410	1.110223e−016	1.879040e−010
0.7	0.6476559551406433	0.6476559551406429	4.440892e−016	1.776835e−010
0.8	0.6648399769821805	0.6648399769821799	5.551115e−016	1.724676e−010
0.9	0.6811859241891134	0.6811859241891132	2.220446e−016	1.847545e−010
1.0	0.6967346701436834	0.6967346701436828	5.551115e−016	3.005770e−010

Table 2. Showing results for growth model problem.

X	Exact Result	Computed Solution	Error in half step method	Error in Sunday *et al.*
0.1	1020.2013400267558	1020.201340026755	0.000000e+000	1.830358e−011
0.2	1040.8107741923882	1040.8107741923882	0.000000e+000	1.250555e−011
0.3	1061.8365465453596	1061.8365465453596	0.000000e+000	1.227818e−011
0.4	1083.2870676749587	1083.2870676749585	2.273737e−013	3.137757e−011
0.5	1105.1709180756477	1105.1709180756475	2.273737e−013	2.216893e−010
0.6	1105.1709180756477	1127.4968515793755	2.273737e−013	2.060005e−010
0.7	1150.2737988572273	1150.2737988572271	2.273737e−013	2.171419e−010
0.8	1173.5108709918102	1173.5108709918102	0.000000e+000	2.216893e−010
0.9	1197.2173631218102	1197.2173631218102	0.000000e+000	2.744400e−010
1.0	1221.4027581601699	1221.4027581601699	0.000000e+000	4.899903e−010

4.3. Problem 3 (Decay Model)

A certain radioactive substance is known to decay at the rate proportional to the amount present. A block of this substance having a mass of 100 g originally is observed. After 40 mins, its mass reduced to 90 g. Find an expression for the mass of the substance at any time and test for the consistency of the block integrator on this problem for $t \in [0,1]$.

The problem has a differential equation of the form;

$$\frac{dN}{dt} = -\alpha N,\ N(0) = 100,\ t \in [0,1] \tag{21}$$

where N represents the mass of the substance at any time t and α is a constant which specifies the rate at which this particular substance decays. Note that,

$$f(0) = 100 \text{ g},\ t = 40 \text{ mins},\ f(40) = 90 \text{ g}$$

Since for any growth/decay problem,

$$f(t) = f(0)e^{\alpha t}$$

$$90 = 100e^{40\alpha}$$

$$\alpha = \frac{(\ln 9 - \ln 10)}{40} = -0.0026$$

Thus, the theoretical solution to (20) is given by,

$$f(t) = 100e^{-0.0026t} \tag{22}$$

which is also the expression for the mass of the substance at any time t.

Applying our new half step numerical scheme (8) to solve the Growth model (22) gives results as shown in **Table 3** [16,17] (**Table 3**).

5. DISCUSSION OF THE RESULT

We have considered three real-life model problems to test the efficiency of our method. Problems 1 and 2 and 3 were solved by Sunday et al. [17]. They proposed an order six block integrator for the solution of first-order ordinary differential equations. Our half-step block method gave better approximation as shown in Tables 1-3 because the iteration per step in the new method was lower than the method proposed by [17]. Our method was found to be zero stable, consistent and converges. **Figure 1** shows the region of absolute stability. From the numerical examples, we could safely conclude that our method gave better accuracy than the existing methods.

Table 3. Showing results for decay model problem.

X	Exact Result	Computed Solution	Error in half step method	Error in Sunday *et al.*
0.1	99.9740033797070850	99.9740033797070850	0.000000e+000	0.000000e+000
0.2	99.9480135176568470	99.9480135176568330	1.421085e−014	1.421085e−014
0.3	99.9220304120923400	99.9220304120923400	0.000000e+000	0.000000e+000
0.4	99.8960540612571460	99.8960540612571460	0.000000e+000	0.000000e+000
0.5	99.8700844633952300	99.8700844633952440	1.421085e−014	0.000000e+000
0.6	99.8441216167510670	99.8441216167510820	1.421085e−014	0.000000e+000
0.7	99.8181655195695610	99.8181655195695750	1.421085e−014	0.000000e+000
0.8	99.7922161700960970	99.7922161700960970	0.000000e+000	0.000000e+000
0.9	99.7662735665764730	99.7662735665764730	0.000000e+000	0.000000e+000
1.0	99.7403377072569700	99.7403377072569700	0.000000e+000	0.000000e+000

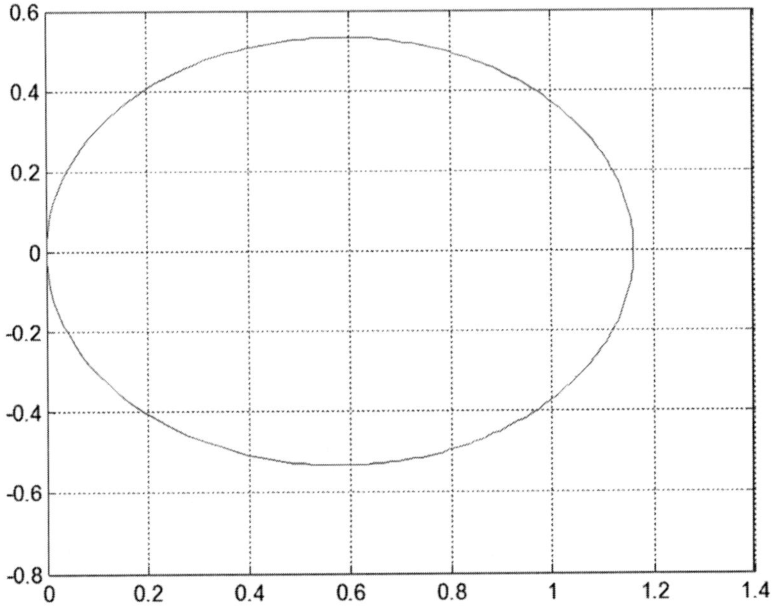

Figure 1. Showing region of absolute stability of our method.

REFERENCES

1. A. O. Adesanya, M. R. Odekunle, and A. A. James, "Order Seven Continuous Hybrid Methods for the Solution of First Order Ordinary Differential Equations," Canadian Journal on Science and Engineering Mathematics, Vol. 3 No. 4, 2012, pp. 154-158.

2. A. M. Badmus and D. W. Mishehia, "Some Uniform Order Block Methods for the Solution of First Ordinary Differential Equation," J. N.A.M. P, Vol. 19, 2011, pp 149- 154.

3. J. Fatokun, P. Onumanyi and U. W. Sirisena, "Solution of First Order System of Ordering Differential Equation by Finite Difference Methods with Arbitrary," J.N.A.M.P, 2011, pp. 30-40.

4. B. I. Zarina and S. I. Kharil, "Block Method for Generalized Multistep Method Adams and Backward Differential Formulae in Solving First Order ODEs," MATHEMATIKA, 2005, pp. 25-33.

5. G. Dahlquist, "A Special Stability Problem for Linear Multistep Methods," BIT, Vol. 3, 1963, pp. 27-43.

6. P. Onumanyi, U. W. Sirisena and S. A. Jator, "Solving Difference Equation," International Journal of Computing Mathematics, Vol. 72, 1999, pp. 15-27.

7. E. A. Areo, R. A. Ademiluyi and P. O. Babatola, "ThreeStep Hybrid Linear Multistep Method for the Solution of First Order Initial Value Problems in Ordinary Differential Equations," J.N.A.M.P, Vol. 19, 2011, pp. 261-266.

8. E. A. Ibijola, Y. Skwame and G. Kumleng, "Formation of Hybrid of Higher Step-Size, through the Continuous Multistep Collocation," American Journal of Scientific and Industrial Research, Vol. 2, 2011, pp. 161-173.

9. H. S. Abbas, "Derivation of a New Block Method Similar to the Block Trapezoidal Rule for the Numerical Solution of First Order IVPs," Science Echoes, Vol. 2, 2006, pp. 10-24.

10. Y. A. Yahaya and G. M. Kimleng, "Continuous of TwoStep Method with Large Region of Absolute Stability," J.N.A.M.P, 2007, pp. 261-268.

11. A. O. Adesanya, M. R. Odekunle and A. A. James, "Starting Hybrid Stomer-Cowell More Accurately by Hybrid Adams Method for the Solution of First Order Ordinary Differential Equation," European Journal of Scientific Research, Vol. 77 No. 4, 2012, pp. 580-588.

12. S. Abbas, "Derivations of New Block Method for the Numerical Solution of First Order IVPs," International Journal of Computer Mathematics, Vol. 64, 1997, p. 325.

13. W. B. Gragg and H. J. Stetter, "Generalized Multistep Predictor-Corrector Methods," Journal of Association of Computing Machines, Vol. 11, No. 2, 1964, pp. 188-209.

14. J. B. Rosser, "A Runge-Kutta Method for All Seasons," SIAM Review, Vol. 9, No. 3, 1967, pp. 417-452.

15. R. E. Mickens, "Nonstandard Finite Difference Models of Differential Equations," World Scientific, Singapore, 1994.

16. O. D. Ogunwale, O. Y. Halid and J. Sunday, "On an Alternative Method of Estimation of Polynomial Regression Models," Australian Journal of Basic and Applied Sciences, Vol. 4, No. 8, 2010, pp. 3585-3590.

17. J. Sunday, M. R. Odekunleans and A. O. Adeyanya, "Or der Six Block Integrator for the Solution of First-Order Ordinary Differential Equations," IJMS, Vol. 3, No. 1 2013, pp. 87-96.

CHAPTER 6

Operational method of solution of linear non-integer ordinary and partial differential equations

K. V. Zhukovsky*

ABSTRACT

We propose operational method with recourse to generalized forms of orthogonal polynomials for solution of a variety of differential equations of mathematical physics. Operational definitions of generalized families of orthogonal polynomials are used in this context. Integral transforms and the operational exponent together with some special functions are also employed in the solutions. The examples of solution of physical problems, related to such problems as the heat propagation in various models, evolutional processes,

Black–Scholes-like equations etc. are demonstrated by the operational technique.

KEYWORDS

Inverse operator; Derivative; Differential equation; Special functions; Hermite and Laguerre polynomials

1. BACKGROUND

Differential equations, besides playing important role in pure mathematics, constitute fundamental part of mathematical description of physical processes. Thus, obtaining the solutions for differential equations is of paramount importance. Few types of differential equations allow explicit and straightforward analytical solutions. The development of computer methods and proper technical means in twenty-first century facilitated equations solving. There are numerous numerical methods for solving differential equations (see, for example, Von Rosenberg 1969; Smith1985; Ghia et al. 1982; Ames 2014; Johnson 2012; Carnahan et al. 1969). However, understanding of the obtained solutions and of the interplay of various parameters in them can be best done in analytical form. Thus, despite the revolutionary breakthrough in numerical calculus, analytical studies remain requested. Exact and approximate solutions are searched, while the first are certainly preferred. Recently some fractional type ordinary and partial differential equations involving non-integer derivatives were explored in Demiray et al. (2015). Exact analytical solutions for such equations were obtained as sums of a vector-type functional (Akinlar and Kurulay 2013). Some partial differential–algebraic equations were also solved by the power series method (Filobello-Nino et al. 2015; Benhammouda and Vazquez-Leal 2014). The solutions were obtained in the form of converging series. Fokker–Planck type equations were recently solved by differential transforms in Hesam et al. (2012); the solutions obtained in form of the form of rapidly convergent series. Semi-analytical techniques for the solution of differential–algebraic equations were developed (Soltanian et al. 2013) and applied for description of an incompressible viscous fluid flow. Approximate solutions for some nonlinear delay differential equations were obtained (Caruntu and Bota 2014) and applied to a biologic model. A modified method of simplest equation was proposed in Vitanov et al. (2015) to find exact analytical solutions of nonlinear partial differential equations. In many cases, these solutions are best formulated in terms of special functions and orthogonal polynomials when used for relevant models of physical processes. Hyperbolic, elliptic Weierstrass and Jacobi type, generalized Airy and Bessel type functions are used (Vitanov et al. 2015; Dattoli et al. 2008, 2009; Appèl and Kampé de Fériet 1926; Dattoli 2000; Dattoli et al. 2005; Zhukovsky 2014, 2015a, b, c,d); expansion in series of Hermite and Laguerre polynomials (Appèl and Kampé de Fériet 1926) are employed. These polynomials possess generalized forms with many variables and indices (Dattoli 2000; Dattoli et al. 2005). In this framework the operational definitions for the polynomials are useful (Erdélyi et al. 1953). The above mentioned recent developments in analytical and semi-analytical equations solutions (Demiray et al. 2015; Akinlar and Kurulay 2013; Filobello-Nino et al. 2015; Benhammouda and Vazquez-Leal 2014; Hesam et al. 2012; Soltanian et al. 2013; Caruntu and Bota 2014; Vitanov et al. 2015) are indeed capable of reducing the size of computational work.

In what follows we shall demonstrate the abilities of the operational approach for solution of differential equations. With the help of this general method we will obtain exact analytical solutions for a broad class of differential equations, including those with non-integer derivatives, evolution type equations, generalized forms of heat, mass transfer and Black–Scholes type equations, involving also the Laguerre derivative operator. Recently, this method was applied for solution of some differential equations in Zhukovsky (2014, 2015a) and Dattoli et al. (2007). These equations cover a broad range of physical problems: from propagation and radiation of accelerated charges to heat and mass transfer (see, for example, Haimo and Markett 1992a, b; Zhukovsky 2016). Operational exponent, employed for solution, finds its application even for description of such fundamentals of nature as quarks and neutrinos (Dattoli and Zhukovsky 2007a, b, 2008). We will not include error analysis in our work since the proposed operational method produces analytical solutions, which satisfy the equations exactly.

When it comes to a numerical analysis, there are also practical and theoretical reasons for examining the process of inverting differential operators. Indeed, the inverse or integral form of a differential equation displays explicitly the input–output relationship of the system. Moreover, integral operators are computationally and theoretically less troublesome than differential operators; for example, differentiation emphasizes data errors, whereas integration averages them. Thus, it may be advantageous to apply computational procedures to differential systems, based on the inverse or integral description of the system.

The evident concept of an inverse function is a function that undoes another function: if an input x into the function f produces an output y, then putting y into the inverse function g produces the output x, and vice versa. i.e.,f(x)=yf(x)=y and g(y)= xg(y)=x or g(f(x))=xg(f(x))=x. If a function f has an inverse f^{-1}, it is invertible and the inverse function is then uniquely determined by f. We can develop similar approach with regard to differential operators. In what follows we will further develop this technique and explore its relation with extended forms of orthogonal polynomials, producing useful relations for solution of a variety of differential equations, by means of inverse derivative. The relevant physical problems will be considered.

Let us denote a common differential operator D=d/dxD=d/dx. The inverse derivative of a function *f*(*x*) is another function *F*(*x*): $D^{-1}f(x)=F(x)D^{-1}f(x)=F(x)$, whose derivative is F′(x)=f(x)F′(x)=f(x). Naturally, we expect anti-derivative or inverse derivative D−1D−1 as the inverse operation of differentiation to be an integral operator. The generalized form of the inverse derivative of *f*(*x*) with respect to *x* evidently is ∫f(x)=F(x)+C∫f(x)=F(x)+C, where *C*—the constant of integration. The action of the inverse derivative operator of the *n*-th order

$$D_x^{-n}f(x)=\frac{1}{(n-1)!}\int_0^x (x-\xi)^{n-1}f(\xi)d\xi, \quad (n\in N=\{1,2,3,\ldots\}) \tag{1}$$

can be complemented with the definition for its zeros order action as follows:

$$D_x^0 f(x)=f(x). \tag{2}$$

Hence, we can write:

$$D_x^{-n}1=\frac{x^n}{n!}, \quad (n\in N_0=N\cup\{0\}). \tag{3}$$

For a general form of differential equation

$$\psi(D)F(x)=f(x), \tag{4}$$

where ψ(D) —differential operator, composed of derivatives or various orders: $D, D^2, \ldots, D^n$ the inverse differential operator $1/\psi(D)$ or $(\psi(D))^{-1}$ is defined, such that

$$\psi(D)(\psi(D))^{-1}f(x) = f(x). \tag{5}$$

From (4) we obtain the particular integral

$$F(x) = (\psi(D))^{-1}f(x) \tag{6}$$

The inverse differential operator $(\psi(D))^{-1}$ is evidently linear, i.e.

$$\frac{1}{\psi(D)}\{af(x) + bg(x)\} = a\frac{1}{\psi(D)}f(x) + b\frac{1}{\psi(D)}g(x), \tag{7}$$

where a and b are constants, $f(x)$ and $g(x)$ are some functions of x. Let us consider an elementary example of the following simple equation:

$$\psi(D)F(x) = e^{\alpha x}, \tag{8}$$

where $\psi(D)$ consists of derivatives of various orders. he action of $\psi(D)$ on $\exp(\alpha x)$ results in $\psi(D)e^{\alpha} = (c_nD + \cdots + c_1D + c_0)e^{\alpha} = \psi(\alpha)e^{\alpha}$; applying inverse operator $(\hat{\psi}(D))^{-1}$ to both sides, we obtain: $e^{ax} = \psi(\alpha)\frac{1}{\psi(D)}e^{\alpha x}$ or $\frac{e^{ax}}{\psi(\alpha)} = \frac{1}{\psi(D)}e^{\alpha x}$ and we conclude that Eq. (8) possesses the following particular integral:

$$F(x) = (\psi(D))^{-1}e^{\alpha x} = \frac{e^{\alpha x}}{\psi(\alpha)}. \tag{9}$$

It can be easily shown by means of the inverse derivative operator that (9), being the solution of Eq. (8), is true also for the operator $\psi(D)\psi(D)$with higher than the second order derivatives. Moreover, it is easy to prove the following identity:

$$(\psi(D))^{-1}e^{\alpha x}f(x) = e^{\alpha x}(\psi(D+\alpha))^{-1}f(x), \tag{10}$$

and the action of the inverse operator $(\psi(D+\alpha))^{-1}(\psi(D+\alpha))^{-1}$ on a function f, which can be expressed via the inverse differential operator (ψ(D))−1(ψ(D))−1, reads as follows:

$$F(x) = (\psi(D+\alpha))^{-1}f(x) = e^{-\alpha x}(\psi(D))^{-1}e^{\alpha x}f(x). \tag{11}$$

2. INVERSE DIFFERENTIAL AND EXPONENTIAL OPERATORS FOR SOLUTION OF SOME NON-INTEGER ORDINARY DIFFERENTIAL EQUATIONS

Let us consider the following equation:

$$\left(\beta^2 - (D+\alpha)^2\right)^{\nu} F(x) = f(x), \quad D+\alpha \equiv \tilde{D}, \tag{12}$$

where we denote operator $\tilde{D} \equiv D+\alpha$ and $\alpha, \beta-$ constants. In order to find the particular integral

$$F(x) = \left(\beta^2 - \tilde{D}^2\right)^{-\nu} f(x) \tag{13}$$

we shall make use of the well-known operational identity (Erdélyi et al. 1953; Srivastava and Manocha 1984), frequently used in fractional derivative calculus:

$$\hat{q}^{-\nu} = \frac{1}{\Gamma(\nu)} \int_0^{\infty} \exp(-\hat{q}t)t^{\nu-1}dt, \quad \min\{\mathrm{Re}(q), \mathrm{Re}(\nu)\} > 0, \tag{14}$$

which reads for the operator $\hat{q} = \beta^2 - \tilde{D}^2$::

$$\left(\beta^2 - \tilde{D}^2\right)^{-\nu} f(x) = \frac{1}{\Gamma(\nu)} \int_0^\infty \exp(-\beta^2 t) t^{\nu-1} \exp(t\tilde{D}^2) f(x) dt. \tag{15}$$

There are several ways to proceed with the solution. One of them consists in the following. Note that differential operator −tD~2−tD~2 in the exponential reduces to the first order derivative with the help of the following integral presentation for the exponential of a square of an operatorp^p^ (Wolf 1979):

$$\exp\left(\hat{p}^2\right) = \frac{1}{\sqrt{\pi}} \int_{-\infty}^{\infty} \exp\left(-\xi^2 + 2\xi\hat{p}\right) d\xi, \tag{16}$$

where $\hat{p} = \sqrt{t}\tilde{D}$ in our case. Thus, the above formula reads as follows:

$$\exp(t\tilde{D}^2) f(x) = \frac{1}{\sqrt{\pi}} \int_{-\infty}^{\infty} \exp(-\xi^2 + 2\xi\sqrt{t}\tilde{D}) f(x) d\xi. \tag{17}$$

Now, if we take into account the action of the operator of translation $\exp(\eta\tilde{D})$ for $\tilde{D} = D + \alpha$:

$$\exp(\eta(D + \alpha)) f(x) = \exp(\eta\alpha) f(x + \eta), \tag{18}$$

the above-sketched operational procedure yields the following expression for the particular integral (13):

$$F(x) = \frac{1}{\sqrt{\pi}\Gamma(\nu)} \int_0^\infty t^{\nu-1} \exp\left((\alpha^2 - \beta^2)t\right) \int_{-\infty}^{\infty} \exp\left(-(\xi - \sqrt{t}\alpha)^2\right) f(x + 2\xi\sqrt{t}) d\xi dt. \tag{19}$$

Upon performing the change of variables, given by

$$\eta = x + 2\xi\sqrt{t} \quad \text{and} \quad t = \tau^2, \tag{20}$$

we finally obtain the solution of the Eq. (12):

$$F(x) = \frac{1}{\sqrt{\pi}\,\Gamma(\nu)} \int_0^\infty \tau^{2(\nu-1)} \exp\left(-(\beta\tau)^2\right) \int_{-\infty}^\infty \exp\left(-\left(\frac{\eta - x}{2\tau}\right)^2 + \alpha(\eta - x)\right) f(\eta)\,d\eta\,d\tau. \tag{21}$$

Equation (12) involves a convolution transform $\phi = \int_{-\infty}^\infty G(x-\eta)f(\eta)d\eta$ with the kernel, equal to the Gauss frequency function $G(x,\tau) = \exp(-(x/2\tau)^2 - \alpha x)$, so that

$$F(x) = \frac{1}{\sqrt{\pi}\,\Gamma(\nu)} \int_0^\infty \tau^{2(\nu-1)} \exp\left(-(\beta\tau)^2\right)\phi(x,\tau)\,d\tau = \frac{1}{\sqrt{\pi}\,\Gamma(\nu)} \int_{-\infty}^\infty e^{2\nu\tau - \beta^2 e^{2\tau}} \phi(x, e^\tau)\,d\tau. \tag{22}$$

Evidently, for $\alpha = 0$

$$\left(\beta^2 - D^2\right)^\nu F(x) = f(x) \tag{23}$$

and its solution becomes the particular case of (15) with the substitution D∼→DD∼→D:

$$F(x) = \frac{1}{\Gamma(\nu)} \int_0^\infty \exp(-\beta^2 t) t^{\nu-1} \hat{S} f(x)\,dt, \tag{24}$$

where the differential operator

$$\hat{S} = \exp(tD_x^2) \equiv \exp(tD^2) \tag{25}$$

was thoroughly explored by Srivastava and Manocha (1984). In particular, for $f(x) = \exp(-x^2)$ we can make use of the Gleisher operational rule (Srivastava and Manocha 1984)

$$\hat{S}f(x) = \exp\left(y\frac{\partial^2}{\partial x^2}\right)\exp(-x^2) = \frac{1}{\sqrt{1+4y}}\exp\left(-\frac{x^2}{1+4y}\right) \tag{26}$$

to obtain the following particular solution:

$$F(x) = \frac{1}{\Gamma(\nu)}\int_0^\infty \frac{\exp(-\beta^2 t)t^{\nu-1}}{\sqrt{1+4t}}\exp\left(-\frac{x^2}{1+4t}\right)dt. \tag{27}$$

The other approach to Eq. (12) consists in combining the exponential operator technique, the inverse derivative formalism and the Gauss transform. Indeed, when solving equations with D+α , we can write the particular integral based upon the operational rule (11), where

$$\psi^{-1}(D) = \left(\beta^2 - D^2\right)^{-\nu}. \tag{28}$$

Proceeding with account for (23), (24) and (25), we compute the result of the action of the operator exp (∂_x^2) on $\exp(\alpha x)g(x)$ with the help of the following chain rule:

$$\exp(y\,\partial_x^2)\exp(\alpha x)g(x) = \exp(\alpha\, x)\exp(\alpha^2 y)\exp(2\alpha\, y\,\partial_x)\exp(y\partial_x^2)g(x), \tag{29}$$

where *y* and *α* are the parameters. It eventually yields the following solution for Eq. (12):

$$F(x) = \frac{1}{\Gamma(\nu)} \int_0^\infty t^{\nu-1} \exp\left\{-(\beta^2 - \alpha^2)t\right\} \hat{\Theta}\hat{S}f(x)dt, \tag{30}$$

where Θ^ is the well-known operator of translation:

$$\hat{\Theta} = \exp(2\alpha t D_x) \equiv \exp\left(2\alpha t \frac{\partial}{\partial x}\right), \quad \hat{\Theta}f(x) = f(x + 2\alpha t) \tag{31}$$

and operator S^S^ is encountered in problems, related to heat propagation and defined in (25). Its action can be written in integral form by means of common Gauss transforms:

$$Fi(x, t) \equiv \hat{S}f(x) = \frac{1}{2\sqrt{\pi t}} \int_{-\infty}^{\infty} \exp\left\{-\frac{(x - \xi)^2}{4t}\right\} f(\xi)d\xi. \tag{32}$$

Thus, we conclude that the integrand of the solution (30) of the Eq. (12), apart the phase and the factor, responsible for the equation dimensionν, is a result of consequent action of operators of heat propagation S^S^ and operator of translation Θ^Θ^ on the function f(x)f(x):

$$F(x, t) \equiv \hat{\Theta}\hat{S}f(x) = Fi(x + 2\alpha t, t) = \frac{1}{2\sqrt{\pi t}} \int_{-\infty}^{\infty} \exp\left\{-\frac{(x + 2\alpha t - \xi)^2}{4t}\right\} f(\xi)d\xi. \tag{33}$$

With these notations we can write the solution as follows:

$$F(x) = \frac{1}{\Gamma(\nu)} \int_0^\infty t^{\nu-1} \exp\left\{-(\beta^2 - \alpha^2)t\right\} F(x,t)dt. \tag{34}$$

The solution is best illustrated by the example of a Gaussian function f(x)f(x):

$$f(x) = \exp(-x^2). \tag{35}$$

With the help of the above described operational procedure and on the account of (26) we easily obtain the following solution of Eq. (12) for $f(x) = \exp(-x^2)$ is given by a Gaussian:

$$F(x) = \frac{1}{\Gamma(\nu)} \int_0^\infty \frac{t^{\nu-1} \exp\{-(\beta^2 - \alpha^2)t\}}{\sqrt{1+4t}} \exp\left\{-\frac{(x+2\alpha t)^2}{1+4t}\right\}dt. \tag{36}$$

So far, we have demonstrated on simple examples how the usage of inverse derivative together with operational formalism and, in particular, with exponential operator technique, provide elegant and easy way to find solutions in some classes of differential equations. In what follows we will apply the concept of inverse differential operator to find solutions of more sophisticated problems, expressed by differential equations.

3. OPERATIONAL APPROACH AND ORTHOGONAL POLYNOMIALS FOR SOLUTION OF SOME NON-INTEGER ORDINARY DIFFERENTIAL EQUATIONS

Despite the traditional presentation of many polynomial families is the expansion in series, they are worth being viewed from operational point of view too. Particularly interesting appears their relation with the exponential operators of derivatives and inverse derivatives and special functions. Recently, Hermite, Laguerre and other polynomial families were reconsidered by means of the operational technique (Dattoli 2000; Dattoli et al. 2005, 2006). Hermite polynomials of two variables are explicitly given by the following operational rule (Dattoli 2000) and the series expansion (Gould and Hopper 1962):

$$H_n^{(m)}(x,y) = \exp\left(y\frac{\partial^m}{\partial x^m}\right)x^n, \quad H_n^{(m)}(x,y) = n!\sum_{r=0}^{[n/m]}\frac{x^{n-mr}y^r}{(n-mr)!r!}. \tag{37}$$

Note, that

$$H_n^{(1)}(x,y) = (x+y)^n \quad \text{and} \quad H_n^{(2)}(x,y) = H_n(x,y), \tag{38}$$

where Hn(x,y)Hn(x,y) are more commonly known Hermite polynomials of two variables

$$H_n(x,y) = \exp\left(y\frac{\partial^2}{\partial x^2}\right)x^n, \quad H_n(x,y) = n!\sum_{r=0}^{[n/2]}\frac{x^{n-2r}y^r}{(n-2r)!r!} \tag{39}$$

with the following generating function:

$$\exp\left(xt + yt^2\right) = \sum_{n=0}^{\infty} \frac{t^n}{n!} H_n(x,y). \tag{40}$$

They can be reduced to the well-known forms of Hermite polynomials of single variable:

$$H_n(x,y) = (-i)^n y^{n/2} H_n\left(\frac{ix}{2\sqrt{y}}\right) = i^n (2y)^{n/2} He_n\left(\frac{x}{i\sqrt{2y}}\right). \tag{41}$$

Note also the following useful and easy to prove relation (Gould and Hopper 1962) for Hermite polynomials:

$$z^n H_n(x,y) = H_n(xz, yz^2). \tag{42}$$

Laguerre polynomials of two variables can be given by an operational relation (Dattoli 2000) or a sum as follows:

$$L_n(x,y) = \exp\left(-y\frac{\partial}{\partial x}x\frac{\partial}{\partial x}\right)\frac{(-x)^n}{n!} = n!\sum_{r=0}^{n}\frac{(-1)^r y^{n-r} x^r}{(n-r)!(r!)^2}. \tag{43}$$

They also reduce to polynomials of a single variable (Srivastava and Manocha 1984) as follows:

$$L_n(x,y) = y^n L_n\left(\frac{x}{y}\right), \quad L_n(x) = y^{-n} L_n(xy,y) = L_n(x,1). \tag{44}$$

The introduction of the second variable in Hermite and Laguerre polynomials allows us to consider them as solutions of partial differential equations with proper initial conditions:

$$\partial_y L_n(x,y) = -(\partial_x x \partial_x) L_n(x,y) \quad \text{with} \quad L_n(x,0) = \frac{(-x)^n}{n!} \tag{45}$$

for Laguerre polynomials Ln(x,y)Ln(x,y) and

$$\partial_y H_n(x,y) = \partial_x^2 H_n(x,y) \quad \text{with} \quad H_n(x,0) = x^n \tag{46}$$

for Hermite polynomials Hn(x,y)Hn(x,y). Importantly, the following differential operators

$${}_L D_x \doteq \frac{\partial}{\partial x} x \frac{\partial}{\partial x} = -\hat{P} \quad \text{and} \quad \hat{M} = y - D_x^{-1} \tag{47}$$

are non commutative:

$$[{}_L D_x, D_x^{-1}] = -1, \quad ([A,B] = AB - BA) \tag{48}$$

and the following operational relation between them exists (Dattoli et al. 2006):

$${}_L D_x = \frac{\partial}{\partial x} x \frac{\partial}{\partial x} = \frac{\partial}{\partial D_x^{-1}}. \tag{49}$$

With the help of this relation, we can extend our approach on differential equations, including operator ∂xx∂x∂xx∂x, sometimes called Laguerre derivative LDx Then, from the definition (47) we immediately conclude for Laguerre polynomials Ln(x,y), defined in

(43), that in terms of inverse derivative operator they are expressed as follows:

$$L_n(x,y) = n!\sum_{k=0}^{n} \frac{(-x)^k y^{n-k}}{(n-k)!(k!)^2} = (y - D_x^{-1})^n\{1\}. \tag{50}$$

Moreover, it follows from (43) and (50) that the following operational identity is true for Laguerre polynomials:

$$\exp\left(\alpha \frac{\partial}{\partial D_x^{-1}}\right) L_n(x,y) = L_n(x, y-\alpha). \tag{51}$$

Various polynomial families, such as Hermite, Laguerre, Legendre, Shaffer and hybrid polynomials can be reviewed in the context of umbral calculus as members of a more general family of Appèl polynomials, which they belong to. Such consideration is possible in the framework operational approach, where inverse derivative plays important role as an instrument for the study of relevant polynomial families, their features and properties.

For example, let us consider Eq. (12) with $f(x) = x^k$. Then, making use of the operational rule (11), and of the identity

$$\exp(yD_x^2)x^k e^{\alpha x} = e^{(\alpha x + \alpha^2 y)} H_k(x + 2\alpha y, y), \tag{52}$$

which arises from the operational relation:

$$\exp\left(y\frac{\partial^m}{\partial x^m}\right) f(x) = f\left(x + my\frac{\partial^{m-1}}{\partial x^{m-1}}\right)\{1\} \tag{53}$$

and from generating function (40), we can write the particular integral (13) for $f(x) = x^k$ as follows:

$$F(x) = \left(\beta^2 - (D_x + \alpha)^2\right)^{-\nu} x^k = \frac{1}{\Gamma(\nu)} \int_0^\infty e^{-t((\beta^2-\alpha^2)} t^{\nu-1} H_k(x + 2\alpha t, t) dt.$$

(54)

The above expression with the shifted argument of the Hermite polynomial can be derived directly from the general form of the solution (30) and the operational definition of the Hermite polynomials (39). Particular solutions for (12) with $f(x) = x^k.$ and $\beta = 0$, obviously follow from (54).

Note, that even without specifying the type of the function f in the r.h.s. of (12) and the values of ν and α and, we can still disentangle two integrals in (21) by involving Hermite polynomials of two variables (40) as follows:

$$F(x) = \frac{1}{\sqrt{\pi}\Gamma(\nu)} \sum_{n=0}^{\infty} \int_0^\infty \tau^{2(\nu-1)} \exp(-\beta^2\tau^2) H_n\left(\alpha, -\frac{1}{4\tau^2}\right) d\tau \frac{1}{n!} \int_{-\infty}^{\infty} (\eta - x)^n f(\eta) d\eta.$$

(55)

Now, let us consider the following equation:

$$\left(\beta - \frac{\partial}{\partial x} x \frac{\partial}{\partial x}\right)^{\nu} F(x) = f(x).$$

(56)

From operational point of view, its solution writes as follows:

$$F(x) = \left(\beta - \frac{\partial}{\partial x} x \frac{\partial}{\partial x}\right)^{-\nu} f(x) = \frac{1}{\Gamma(\nu)} \int_0^\infty \exp(-\beta t) t^{\nu-1} \exp(t_L D_x) f(x) dt.$$

(57)

Note, that upon the change of variable t→et the last integral can be transformed into the following

$$F(x) = \left(\beta - \frac{\partial}{\partial x} x \frac{\partial}{\partial x}\right)^{-\nu} f(x) = \frac{1}{\Gamma(\nu)} \int_{-\infty}^{\infty} e^{t\nu} e^{-\beta e^t} e^{e^t {}_L D_x} f(x) dt.$$

(58)

In the particular case of β=1,ν=1β=1,ν=1 the above integral presentation reduces to the Laplace transforms for the differential operator LDxLDx, involved in (57), identical, except for the change LDx↔Dx, to the well known Laplace transforms for the operator Dx

$$\frac{1}{1-\hat{D}_x} = \int_0^{\infty} \exp(-s(1-\hat{D}_x))ds. \tag{59}$$

By choosing $f(x) = x^n$ in Eq. (57), we obtain the following result:

$$(\beta - {}_L D_x)^{-\nu} x^n = \frac{n!}{\Gamma(\nu)} \int_0^{\infty} \exp(-\beta t) t^{n+\nu-1} L_n(x/t) dt. \tag{60}$$

Moreover, we can write a solution of Eq. (56) for an arbitrary function f(x)f(x), if its expansion into series of simple Laguerre polynomials Ln(x)Ln(x)exists. Then, provided

$$f(x) = \sum_{n=0}^{\infty} c_n L_n(x), \tag{61}$$

and taking into account (51), the solution (57) of Eq. (56) can be also written as the following integral of series of Laguerre polynomials:

$$F(x) = \frac{1}{\Gamma(\nu)} \int_0^\infty \exp(-\beta t) t^{\nu-1} \sum_{n=0}^\infty c_n L_n(x, 1-t) dt. \tag{62}$$

For the exponential function $f(x) = \exp(-\gamma x)$ we can employ the generalized form of the Gleisher operational rule (see Dattoli et al. 2007)

$$\exp(-t_L D_x) \cdot \exp(-\gamma x) = \frac{1}{1-\gamma t} \exp\left(-\frac{\gamma x}{1-\gamma t}\right), \tag{63}$$

which immediately yields the following result:

$$\left(\beta - \frac{\partial}{\partial x} x \frac{\partial}{\partial x}\right)^{-\nu} \exp(-\gamma x) = \frac{1}{\Gamma(\nu)} \int_0^\infty \exp(-\beta t) t^{\nu-1} \frac{1}{1+\gamma t} \exp\left(-\frac{\gamma x}{1+\gamma t}\right) dt. \tag{64}$$

Another interesting case arises in the case of the Laguerre derivative LDxLDx instead of the common derivative ∂x∂x in the Eq. (12). Let us choose the following initial condition function:

$$f(x) = W_0(-x^2, 2), \tag{65}$$

where

$$W_n(x, m) = \sum_s \frac{x^s}{s!(ms+n)!}, \tag{66}$$

$W_n(x, m)$— particular case of the Bessel–Write function (Srivastava and Manocha 1984). Then, following (14), we can write the solution:

$$\left(\beta^2 - \left(\frac{\partial}{\partial x} x \frac{\partial}{\partial x}\right)^2\right)^{-\nu} W_0(-x^2, 2) = \frac{1}{\Gamma(\nu)} \int_0^\infty \exp(-\beta^2 t) t^{\nu-1} \exp(t_L D_x^2) f(x) dt. \tag{67}$$

Eventually, based on the operational definition of Laguerre polynomials (43) and exploiting the other generalized form of the Gleisher operational rule (Dattoli et al. 2006) in the form

$$\exp({}_L D_x^2) W_0(-x^2, 2) = \frac{1}{\sqrt{1+4t}} W_0\left(-\frac{1}{1+4t}, 2\right), \tag{68}$$

we obtain:

$$\left(\beta^2 - \left(\frac{\partial}{\partial x} x \frac{\partial}{\partial x}\right)^2\right)^{-\nu} W_0(-x^2, 2) = \frac{1}{\Gamma(\nu)} \int_0^\infty \exp(-\beta^2 t) t^{\nu-1} \frac{1}{\sqrt{1+4t}} W_0\left(-\frac{1}{1+4t}, 2\right) dt. \tag{69}$$

Moreover, according to the developed above procedure, we can write the solutions for other than above specified types of equations. For example, we can take advantage of the following generalization of the Laguerre polynomials $L_n^{(\alpha)}(x, y)$:

$$L_n^{(\alpha)}(x, y) = \exp\left[-y\breve{D}_x\right]\left\{\frac{(-x)^n}{n!}\right\}, \tag{70}$$

where operator $\breve{D}_x$ is defined as follows:

$$\breve{D}_x = x\partial_x^2 + (\alpha+1)\partial_x. \tag{71}$$

Inverse operator technique easily allows us to write the solution of the equation

$$(x\partial_x^2 + (\alpha + 1)\partial_x)^{\nu} F(x) = f(x). \tag{72}$$

Indeed, by following the operational rule (14), we get

$$\breve{D}_x^{-\nu} f(x) = \frac{1}{\Gamma(\nu)} \int_0^{\infty} \exp(-\beta t) t^{\nu-1} \exp(t\breve{D}_x) f(x) dt \tag{73}$$

and for initial the condition function $f(x) = x^n$, we then write:

$$\breve{D}_x^{-\nu} x^n = \frac{(-1)^n n!}{\Gamma(\nu)} \int_0^{\infty} \exp(-\beta t) t^{\nu-1} L_n^{(\alpha)}(x, -t) dt. \tag{74}$$

Similarly to (60), taking advantage of the generalized form of the Gleisher operational rule from Srivastava and Manocha (1984), we obtain for the operator D⁀xD⁀x and $f(x) = \exp(-\gamma x)$: :

$$\breve{D}_x^{-\nu} \exp(-\gamma x) = \frac{1}{\Gamma(\nu)} \int_0^{\infty} \exp(-\beta t) t^{\nu-1} \frac{1}{(1 + \gamma t)^{\alpha+1}} \exp\left(-\frac{\gamma x}{1 + \gamma t}\right) dt. \tag{75}$$

4. OPERATIONAL SOLUTION OF SOME PARTIAL DIFFERENTIAL EQUATIONS

The method of the inverse differential operators has multiple applications for solving mathematical problem, describing wide range of physical processes, such as the heat transfer, the diffusion, wave propagation etc. Some of the examples of solution of the heat equation, of the diffusion equation and of their modified forms, the Laguerre heat equation and others, by the inverse derivative method were considered in Dattoli et al. (2006, 2007) and Zhukovsky and Dattoli (2011). In what follows we will explore more complicated, generalized forms of the aforementioned equations, as well as some second order over the time variable partial differential equations will be touched on.

It is worth mentioning that, despite the relation (11) seems trivial to all appearance, it is very useful for solution of a broad family of diferential equations by operational method.

Indeed, for the diferential equation $\psi(D_x+\alpha)F(x,t)=f(x,t)$ we can rewrite (11) in the following form: $e^{\alpha x}F(x,t)=\psi^{-1}(D_x)e^{\alpha x}f(x,t)$ and, for example, for the evolutional type equations, where $f(x,t)=\partial_t F(x,t)$, we obtain $\psi(D_x)e^{\alpha}F(x,t)=\partial_t e^{\alpha}F(x,t)$. By denoting $e^{\alpha}F(x,t)=G(x,t)$ we have the equation $\psi(D_x)G(x,t)=\partial_t G(x,t)$ with $\psi(D_x)$ and with the initial condition $g(x)=G(x,0)=e^{\alpha x}F(x,0)=e^{\alpha x}f(x)$. hus, in order to obtain the desired solution $F(x,t)=e^{\alpha}G(x,t)$ of the equation $\psi(D_x+\alpha)F(x,t)=f(x,t)$ with the initial condition $F(x,0)=f(x)$, we end up with the necessity to solve the equation with $\psi(D_x)$ for the function $G(x,t)$ with the initial condition $g(x)=e^{\alpha x}f(x)$. Note that the above discussed method is applicable not only to the evolutional type equations with ∂_t in the r.h.s., but also to other operators $\hat{D}(t)$, act- ing over the time variable. Indeed, if $G(x,t)$ is the solution of $\psi(D_x)G(x,t)=\hat{D}(t)G(x,t)$ with $g(x)=$

$G(x,0) = e^{\alpha x}F(x,0) = e^{\alpha x}f(x)$, then, following the above scheme, it is easy to demonstrate that $F(x, t) = e^{\alpha}\ G(x, t)$ is the solution of the equation $\psi(D_X + \alpha)F(x, t) = \hat{D}(t)F(x, t)$ with $F(x, 0) = f(x)$. Evidently, in the case of the second order diferential operator $\hat{D}(t)$ the second boundary or initial condition has to be chosen for the diferential equation for F(x, t) and, accordingly, for G(x, t). In what follows, we shall apply the above-discussed method to several examples of equations, common in physics and not only.

5. BLACK–SCHOLES TYPE EQUATIONS

To demonstrate the solution of differential equations by the operational method we first consider the following differential equation, which is a generalized form of a Black–Scholes equation, frequently used in financial models:

$$\frac{1}{\rho}\frac{\partial}{\partial t}F(x,t) = \left[x^2\frac{\partial^2}{\partial x^2} + 2\alpha x^2\frac{\partial}{\partial x} + \lambda x\frac{\partial}{\partial x} + (\alpha x)^2 - \mu\right]F(x,t), \quad f(x) = F(x,0), \tag{76}$$

where α, ρ, λ andи μ are the constant coefficients and f(x)= F(x,0)f(x)=F(x,0) is the initial condition function. The apparently complicated Eq. (76) reduces to the following form:

$$\frac{1}{\rho}\frac{\partial}{\partial t}G(x,t) = x^2\frac{\partial^2}{\partial x^2}G(x,t) + \lambda x\frac{\partial}{\partial x}G(x,t) - \mu G(x,t), \quad g(x) = G(x,0), \tag{77}$$

by the substitution $\partial x \to \partial x + \alpha \partial x \to \partial x + \alpha$. Therefore, according to (11) and to the discussion in the beginning of this chapter, the solution of the Eq. (76) will be found, if we obtain the solution of the Black–

Scholes Eq. (77) for G(x,t)G(x,t) with the initial condition function $g(x) = G(x,0) = e^{\alpha x} F(x,0)$. Then, the solution of (76) reads as follows:

$$F(x,t) = e^{-\alpha x} G(x,t), \quad \text{with} \quad g(x) = G(x,0) = e^{\alpha x} F(x,0). \tag{78}$$

The Eq. (77) can be easily solved with the help of the operational approach if we distinguish the perfect square of the operator x∂xx∂x (see Dattoli et al. 2007):

$$G(x,t) = \frac{\exp(-\rho\varepsilon t)}{\sqrt{\pi}} \int_{-\infty}^{\infty} \exp\left[-\sigma^2 + \sigma\gamma\frac{\lambda}{2\rho}\right] g(x\exp(\sigma\gamma))d\sigma, \tag{79}$$

where $\gamma = \gamma(t) = 2\sqrt{\rho t}, \quad \varepsilon = \mu + (\lambda/2)^2$. Let us choose the initial condition $f(x) = e^{-\alpha x} x^n$ for the Black–Scholes type Eq. (76), i.e. $g(x) = x^n$; then the solution (79) has the simple form $G(x,t) = x^n \exp\{\rho t(n^2 + \lambda n - \mu)\}$ (see Dattoli et al. 2007) and the Eq. (76) has the following solution:

$$F(x,t) = e^{-\alpha x} x^n \exp\{\rho t(n^2 + \lambda n - \mu)\}. \tag{80}$$

Let us consider another generalization of the Black–Scholes type differential equation with the Laguerre derivative (see 43, 47):

$$\frac{1}{\rho}\frac{\partial}{\partial t} A(x,t) = (\partial_x x \partial_x)^2 A(x,t) + \lambda(\partial_x x \partial_x) A(x,t) - \mu A(x,t), \quad g(x) = A(x,0), \tag{81}$$

where ρ, λ and μ are the constant coefficients and g(x)= A(x,t=0)g(x)=A(x,t=0) is the initial condition function. The Eq. (81)

generalizes and unifies equations of Laguerre diffusion of matter and of heat, considered in Dattoli et al. (2005, 2007). This equation also can be solved by the operational method developed above. Indeed, by distinguishing the perfect square of the Laguerre derivative LDx=∂xx∂xLDx=∂xx∂x in (81), the solution evidently reads in the form of the exponential $A(x,t)$ = exp ρt (LDx + /2)^2 – ε $g(x)$, where ε = μ + (/2)^2 Now we apply the operational identity (16) to exp($aLDx$) to obtain the following solution for A(x, t):

$$A(x,t) = \frac{\exp(-\varepsilon\alpha^2)}{\sqrt{\pi}} \int_{-\infty}^{\infty} \exp\left(-\sigma^2 - \sigma\alpha\lambda - 2\sigma\alpha_L D_x\right) g(x) d\sigma, \tag{82}$$

where $\alpha = \alpha(t) = \sqrt{\rho t}$. For a given function g(x) we still have to ind the result of the result of the action of the operational exponent exp(−$aLDx$)g (x) and to take the integral dσ. Let us choose, for example, the initial condition function g(x) = (−x)^n/(−x)^{n}n!. hen we can make use of the operational deinition of the Laguerre polynomials (43) and obtain the integral form for A

$$A(x,t) = \frac{\exp(-\varepsilon\alpha^2/4)}{\sqrt{\pi}} \int_{-\infty}^{\infty} \exp\left(-\sigma^2 - \sigma\alpha\lambda\right) L_n(x, 2\sigma\alpha) d\sigma. \tag{83}$$

Further integration over dσdσ yields the solution of the Black–Scholes equation with the Laguerre derivative (81) and with the initial condition $g(x) = (-x)^n/n!$ in the following form:

$$A(x,t) = \frac{\exp(-\alpha^2\mu)}{\sqrt{\pi}} n! \sum_{r=0}^{n} \frac{(-x)^r (2\alpha)^{n-r}}{(n-r)!(r!)^2} I, \tag{84}$$

where

$$I = \frac{\alpha\hat{\lambda}}{2}\left(e^{i(n-r)\pi} - 1\right)\Gamma\left(1 + \frac{n-r}{2}\right){}_1F_1\left(\frac{1-(n-r)}{2}, \frac{3}{2}, -\left(\frac{\alpha\hat{\lambda}}{2}\right)^2\right)$$
$$+ \frac{1}{2}\left(e^{i(n-r)\pi} + 1\right)\Gamma\left(\frac{1+n-r}{2}\right){}_1F_1\left(-\frac{n-r}{2}, \frac{1}{2}, -\left(\frac{\alpha\hat{\lambda}}{2}\right)^2\right),$$

(85)

ΓΓ is the gamma function and 1F11F1 is the hypergeometric function. Evidently, if the initial condition function can be expanded in the power series of xx, then the respective solution represents series of the already obtained solution (84). Moreover, if the expansion in series of the Laguerre polynomials for the initial condition function $g(x) = \sum_n a_n L_n(x)$ exists, then we can exploit the relationships (51) and (44) to obtain the solution in the following form:

$$A(x,t) = \frac{\exp(-\varepsilon\alpha^2/4)}{\sqrt{\pi}} \sum_n a_n \int_{-\infty}^{\infty} \exp\left(-\sigma^2 - \sigma\alpha\hat{\lambda}\right) L_n(x, 2\sigma\alpha + 1) d\sigma.$$

(86)

In the most general case the solution A(x,t)A(x,t) can be obtained through the following procedure: we employ the operational definitions (47) and the definition of the inverse derivative (1) to write

$$A(x,t) = \frac{\exp(-\varepsilon\alpha^2)}{\sqrt{\pi}} \int_{-\infty}^{\infty} \exp\left(-\sigma^2 - \sigma\alpha\hat{\lambda}\right) \exp\left(-2\sigma\alpha \frac{\partial}{\partial D_x^{-1}}\right) \phi(D_x^{-1}) 1 d\sigma,$$ (87)

where $\phi(D_x^{-1})\,\mathbf{1} = g(x)$ and the explicit form of the function ϕϕ is given by the integral $\phi(x) = \int_0^\infty \exp(-\kappa) g(x\kappa) d\kappa$, provided it converges. Note, that $(-t_L D_x)g(x)$ is the solution of the Laguerre diffusion equation (Dattoli et al. 2007) $\partial_t f(x,t) = -_L D_x f(x,t)$ with the initial condition f(x,0)=g(x); therefore, the result of the action of the exponential operator in (87) is in fact $f(x,t) = \exp\left(-t\frac{\partial}{\partial D_x^{-1}}\right)g(x) = \phi(D_x^{-1} - t)1-$ the solution of the above mentioned Laguerre diffusion equation. Then the desired solution of the Eq. (81) takes the following form:

$$A(x,t) = \frac{\exp(-\varepsilon\alpha^2)}{\sqrt{\pi}} \int_{-\infty}^{\infty} \exp\left(-\sigma^2 - \sigma\alpha\lambda\right)g(x,t)d\sigma, \tag{88}$$

where

$$g(x,t) = \phi(D_x^{-1} - 2\sigma\alpha)\mathbf{1} = \exp\left(-2\sigma\alpha\frac{\partial}{\partial D_x^{-1}}\right)\phi(D_x^{-1})\mathbf{1}. \tag{89}$$

Consider the following initial condition: $g(x) = W_0(-x^2, 2)$, $W_n(x,m) = \sum_{r=0}^{\infty} \frac{x^r}{r!(mr+n)!}$, is the particular case of the Bessel–Write function (Srivastava and Manocha 1984). The corresponding image function is $\phi(x) = \exp(-x^2)$. With account for (16) and (89) we obtain (see also Dattoli et al. 2007)

$$g(x,t) = \frac{1}{\sqrt{\pi}} \int_{-\infty}^{\infty} \exp\left(-\xi^2 + 4i\sigma\alpha\xi\right)C_0(2i\xi x)d\xi, \tag{90}$$

where $C_n(x) = \sum_{r=0}^{\infty} \frac{(-x)^r}{r!(r+n)!}$, $(n \in \mathrm{N}_0)$, is the Bessel–Tricomi function (Watson 1944), related to the Bessel–Write function $C_n(x) = W_n(-x, 1)$ and to the commonly known cylindric Bessel functions $C_n(x) = x^{-n/2}J_n(2\sqrt{x})$. Thus, the solution (81) with the initial condition $g(x) = W_0(-x^2, 2)$ is explicitly determined by the formulae (88) and (90).

6. HEAT DIFFUSION TYPE EQUATIONS

Let us consider the following generalized heat type equation with the linear term

$$\partial_t F(x,t) = \alpha \partial_x^2 F(x,t) + \beta x F(x,t) \tag{91}$$

with the initial condition $F(x,0) = f(x)$. The solution of Eq. (91) reads (for example, see Zhukovsky and Dattoli 2011 for α=1α=1):

$$F(x,t) = e^{\Phi(x,t;\beta)} \hat{\Theta} \hat{S} f(x) = e^{\Phi(x,t;\beta)} f(x + \beta t^2, t), \tag{92}$$

where $\hat{\Theta} = e^{ab\partial_x}$, $\hat{S} = e^{a\,\partial_x^2}$, $\Phi(x,t;\beta) = \frac{1}{3}ab^2 + bx$, $a = \alpha t$, $b = \beta t$. The solution (92) consists in the action of the evolution operator on the initial condition $F(x,0) = f(x)$, which is transformed by $\hat{S}$ and $\hat{\Theta}$. Let us choose now the initial condition $f(x) = x^n$ for the heat diffusion type Eq. (91). Then, upon the action of the S¯^ operator on it and according to the operational definition of the Hermite polynomials (39) $e^{a\partial_x} x^n = H_n(x,a)$, we obtain the solution $F(x) \propto H_n(x + ab, a)$, and we end up with

$$F(x,t) = e^{\Phi} H_n\left(x + \alpha\beta t^2, \alpha t\right). \tag{93}$$

Let us now consider the following equation:

$$\partial_t F(x,t) = \partial_x^2 F(x,t) + 2\delta\partial_x F(x,t) + \beta x F(x,t) + \gamma F(x,t), \quad F(x,0) = f(x). \tag{94}$$

It can be viewed as Eq. (91) where we substituted ∂x→∂x+δ and set α=1α=1. Distinguishing the perfect square of the operator ∂x+δ we can search the desired solution of the Eq. (94) in the following form:

$$F(x,t) = \exp\left(t\left(\gamma - \delta^2\right) - \delta x\right) G(x,t), \tag{95}$$

where $G(x,t)$ satisies the Eq. (91) for G with the initial condition $G(x,0) = g(x)$, g(x) = exp(δx)f (x). Let us choose the initial condition for (94), for example, in the form of the powers of x: f (x) = x^k. hen $g\ (x) = x^k e^{\delta x}$ and we have in fact Eq. (91) for: $\partial_t G = (\partial_x^2 + \beta x)G$. Upon the action of the $\hat{\hat{S}}$ operator on it and due to the operational rule (52) we obtain $\hat{S}g(x) = \exp(\delta(x+\delta a))H_k(x+2\delta a, a) = g(x,t)$. he consequent action of the translation operator $\hat{\Theta}$ yields the shift along the x argument and, thus, the solution of Eq. (91) for G, taken $G(x,0) = g(x)$, has the following form:

$$G(x,t) = e^{\Phi+\Delta_1} H_k\left(x + \left(\alpha\delta^2/\beta\right)\left(2t\beta/\delta + t^2(\beta/\delta)^2\right), \alpha t\right), \tag{96}$$

where $\Delta 1 = \delta(x + \delta\alpha t + \alpha\beta t^2)$ For δ=0 it immediately returns the result (93). Studying the evolution at prolonged times, such that t>>δ/β notice that Δ1<<Φ and, provided at so long times the special condition x>>δ²α/β is also fulfilled, that is the coordinate travel is limited, we end up with the separation of the dependence of the solution on time and coordinate. The coordinate dependence is contained in the exponential factor exp(βtx)exp(βtx), while the time is contained in this factor as well as in the Hermite polynomial arguments Hk(2δαt+αβt2,αt)Hk(2δαt+αβt2,αt). For the short times of the evolution of the system, such that t<<δ/β the phase approximately reads Φ+Δ1≅xδ+αδ2t and the Hermite polynomials depend on both coordinate and time: Hk(x,αt) Thus, for relatively short times we have the solution approximated by $G(x,t)|_{t\to 0} = e^{x\delta+\alpha\delta^2 t} H_k(x,\alpha t)$ and for infinitely short times αt→0αt→0 the Hermite polynomials become $H_k(x,0) = x^k$, which is perfect agreement with our initial condition $g(x) = x^k e^{\delta x}$. The desired solution of the Eq. (94) follows immediately upon the assumption of α=1α=1 in (96) with different phase $\Delta_2 = t\gamma + t^2\delta\beta$:

$$F(x,t) = e^{\Phi+\Delta_2} H_k\left(x + 2t\delta + t^2\beta, t\right). \tag{97}$$

The two-dimensional heat diffusion type equation with the linear terms

$$\partial_t F(x,y,t) = \left\{\left(\alpha\partial_x^2 + \beta\partial_x\partial_y + \gamma\partial_y^2\right) + bx + cy\right\}F(x,y,t), \quad \min(\alpha,\beta,\gamma) > 0, \tag{98}$$

and the initial condition $F(x,y,0) = f(x,y)$ can be solved following (Zhukovsky 2014) or with the help of the Baker–Campbell–Hausdorf formula exp $\left(\hat{A}+\hat{B}\right) = \exp\left(-[\hat{A},\hat{B}]/2\right)\exp\left(\hat{A}\right)\exp\left(\hat{B}\right)$. In complete analogy with the one-dimensional case, we obtain the solution of the two-dimensional heat conduction equation with lineal terms (98) in the following form:

$$F(x,y,t) = e^{\Psi}\hat{\Theta}_x\hat{\Theta}_y\hat{E}f(x,y) \propto f\left(x + t^2(\alpha b + \beta c/2), y + t^2(\gamma c + \beta b/2), t\right). \tag{99}$$

where $\Psi = (\alpha b^2 + \gamma c^2 + \beta bc)t^3/3 + t(bx + cy)$ is the phase, $\hat{\Theta}_x = e^{t^2(\alpha b + \beta c/2)\partial_x}$ and $\hat{\Theta}_y = e^{t^2(\gamma c + \beta b/2)\partial_y}$ re the diffusion operators for each of the two coordinates, and

$$\hat{E} = \exp\left[t\left(\alpha\partial_x^2 + \beta\partial_x\partial_y + \gamma\partial_y^2\right)\right] \tag{100}$$

is the two-dimensional analogue of heat diffusion operator Ŝ (25). The explicit double integral form of the operator Ê was obtained in Dattoli et al. (2006) and it is the Gauss type integral, which we omit here for the sake of conciseness. It is easy to demonstrate that in the case of β = 0 the heat diffusion is executed by the one-dimensional operators (25) $\hat{S}x\hat{S}y$ instead of the more general operator Ê The solution in this case reads as follows:

$$F(x,y,t) = e^{\Psi}\hat{\Theta}_x\hat{\Theta}_y\hat{S}_x\hat{S}_yf(x,y) \propto f\left(x + t^2\alpha b, y + t^2\gamma c, t\right). \tag{101}$$

Note, that the operational definition (100) allows for the direct computation of the result in many cases, avoiding the necessity to calculate the double integral for the operator E^E^. Let us, for example, choose the initial function in the form of the powers of *x* and *y*: $f(x,y) = x^m y^n$. Then, according to the operational definition of the Hermite polynomials of four variables and two indices Hm,n(x,tα,y,tγ|β) (see, for example, Erdélyi et al. 1953; Dattoli et al. 2006, 2007) we obtain

$$\hat{\mathrm{E}}\{x^m y^n\} = H_{m,n}(x, t\alpha, y, t\gamma | t\beta), \tag{102}$$

where $H_{m,n}$(x,tα,y,tγ|β) are the above mentioned Hermite polynomials with the following generating exponent:

$$\sum_{m,n}^{\infty} \frac{u^m v^n}{m!n!} H_{m,n}(x, \alpha, y, \gamma | \beta) = \exp\left(xu + \alpha u^2 + yv + \gamma v^2 + \beta uv\right). \tag{103}$$

The presentation of Hm,n(x,tα,y,tγ|β) in the form of sums (see Erdélyi et al. 1953) of the two-variable Hermite polynomials Hm(x,y)Hm(x,y), defined in (39), reads as follows:

$$H_{m,n}(x, \alpha; y, \gamma | \beta) = m!n! \sum_{s=0}^{\min(m,n)} \frac{\beta^s}{s!(m-s)!(n-s)!} H_{m-s}(x, \alpha) H_{n-s}(y, \gamma). \tag{104}$$

The action of the translation operators $\hat{\Theta}_x \hat{\Theta}_y$ on the Hermite polynomials $H_{m,n}$(x,tα,y,tγ|β) yields the solution of the two-dimensional heat type equation with the linear terms (98) and with the initial condition in the form of powers $f(x,y) = x^m y^n$:

$$F(x,t) = e^{\Psi} H_{m,n}\left(x + t^2(\alpha b + \beta c/2), t\alpha : y + t^2(\gamma c + \beta b/2), t\gamma \middle| t\beta\right). \tag{105}$$

It appears evident that the obtained solution (105) of the two-dimensional heat conduction problem (98) represents a direct generalization of the solution (93) for the one-dimensional heat conduction analog (91).

7. OPERATIONAL SOLUTION OF SOME SECOND ORDER OF TIME PARTIAL DIFFERENTIAL EQUATIONS

The operational method for solution of differential equations can be successfully applied for the second order of time partial differential equations too. Let us consider the equations of the following type:

$$\left(\frac{\partial^2}{\partial t^2} + \varepsilon \frac{\partial}{\partial t}\right) F(x,t) = \hat{D}(x) F(x,t), \tag{106}$$

where $\hat{D}(x)$ is a differential operator, acting over the coordinate, such as, for example, the heat diffusion operator ∂_x^2 or the Laguerre derivative ${}_L D_x$ or any other. General solution of the Eq. (106) reads as follows:

$$F(x,t) = e^{-\frac{t\varepsilon}{2}}\left(e^{-\frac{t}{2}\sqrt{\varepsilon^2+4\hat{D}(x)}}C_1(x) + e^{\frac{t}{2}\sqrt{\varepsilon^2+4\hat{D}(x)}}C_2(x)\right), \tag{107}$$

where $C_{1,2}(x)$ are to be determined from the initial conditions. Suppose the initial condition F(x,0)=f(x) is given and, for example, we can require for the second order Eq. (106) that its solution converges at infinite time t→∞t→∞. Other initial and boundary conditions are, of course, possible, but they will be considered elsewhere. Our choice sets C2(x)=0C2(x)=0 and the remaining branch of the solution is subject to the Laplace transforms

$$e^{-t\sqrt{V}} = \frac{t}{2\sqrt{\pi}}\int_0^\infty \frac{d\xi}{\xi\sqrt{\xi}}e^{-\frac{t^2}{4\xi}-\xi V}, \quad t>0. \tag{108}$$

Thus, we obtain for the Eq. (106) the following fading at infinite times solution:

$$F(x,t) = e^{-\frac{\varepsilon t}{2}}\frac{t}{4\sqrt{\pi}}\int_0^\infty \frac{d\xi}{\xi\sqrt{\xi}}e^{-\frac{t^2}{4\xi}-\xi\varepsilon^2}\exp\left(-4\xi\hat{D}(x)\right)f(x), \tag{109}$$

provided the integral converges. The particular form of the solution depends on the operator $\hat{D}(x)$ and on the initial function $f(x)$.

Let us first consider the following equation:

$$\frac{\partial^2}{\partial t^2}F(x,t) = \hat{D}(x)F(x,t), \quad F(x,0) = f(x), F(x,\infty) < \infty. \tag{110}$$

Then (107) reduces to $(x,t) = \left(e^{-t\sqrt{\hat{D}(x)}}C_1(x) + e^{t\sqrt{\hat{D}(x)}}C_2(x)\right)$. For $\hat{D}(x) = {}_L D_x$ it becomes

$$\frac{\partial^2}{\partial t^2}F(x,t) = \partial_x x \partial_x F(x,t). \tag{111}$$

We now make use of the identity

$$\frac{t}{2}\int_0^{\infty} \frac{u^m du}{u\sqrt{u}} \exp\left(-\frac{t^2}{4u}\right) = 4^{-m} t^{2m} \Gamma\left(\frac{1}{2} - m\right), \tag{112}$$

which yields the following bounded at infinite times solution of Eq. (111) for the initial function $f(x) = x^n$ in terms of the Gegenbauer polynomial C_n^k and of the gamma function Γ:

$$F(x,t) = \frac{n!}{\sqrt{\pi}} \sum_{r=0}^{n} \frac{(-x)^r 4^{-(n-r)} t^{2(n-r)} \Gamma[1/2 - (n-r)]}{(n-r)!(r!)^2} = n! \frac{(-x)^n C_{2n}^{(-2n)}\left(\frac{t}{2\sqrt{x}}\right)}{\Gamma(1+2n)}. \tag{113}$$

For example, for m=2m=2 (113) immediately reduces to $F(x,t)|_{n=2} = \left(t^4 + 12t^2 x + 6x^2\right)/12.$

Let us consider the Black–Scholes type differential operator in the r.h.s. of the second order differential equation, namely

$$\left(\frac{\partial^2}{\partial t^2}+\varepsilon\frac{\partial}{\partial t}\right)G(x,t)=\left(x^2\frac{\partial^2}{\partial x^2}+\lambda x\frac{\partial}{\partial x}-\mu\right)G(x,t),\quad G(x,0)=g(x),G(x,\infty)<\infty. \tag{114}$$

Distinguishing the full square of the operator $\bar{D}=x\partial_x$, we have to compute the result of the action of the exponential operator exp $\left[-4\xi\left(\left(\bar{D}+\lambda/2\right)^2-\nu\right)\right]g(x)$, where $\nu=\mu+(\lambda/2)^2$. To achieve it, we exploit the operational identity (16), applying it to the exponential exp $\left[(\bar{D}+\lambda/2)^2\right]$:. $e^{-4\xi(\bar{D}+\lambda/2)^2}=\int_{-\infty}^{\infty}\exp\left[-u^2+2iu\lambda\sqrt{\xi}+4iu\sqrt{\xi}\bar{D}\right]du/\sqrt{\pi}$. Upon the action on g(x)g(x), we obtainexp $\left[4iu\sqrt{\xi}x\partial_x\right]g(x)=g\left(e^{4iu\sqrt{\xi}}x\right)$, which yields the function

$$g(x,\xi)=e^{-4\xi(\bar{D}+\lambda/2)^2}f(x)=\frac{1}{\sqrt{\pi}}\int_{-\infty}^{\infty}e^{-u^2+2iu\lambda\sqrt{\xi}}g\left(e^{4iu\sqrt{\xi}}x\right)du, \tag{115}$$

and the solution of Eq. (114) now takes the following form:

$$G(x,t)=e^{-t\varepsilon/2}\frac{t}{4\sqrt{\pi}}\int_0^{\infty}\frac{d\xi}{\xi\sqrt{\xi}}\exp\left[-\frac{t^2}{16\xi}-\varepsilon^2\xi+4\xi\nu\right]g(x,\xi). \tag{116}$$

Eventually, let us consider the following rather complicated differential equation of second order of time and coordinate:

$$\left(\frac{\partial^2}{\partial t^2}+\varepsilon\frac{\partial}{\partial t}\right)F(x,t)=\left[x^2\frac{\partial^2}{\partial x^2}+2\alpha x^2\frac{\partial}{\partial x}+\lambda x\frac{\partial}{\partial x}+(\alpha x)^2-\mu\right]F(x,t).$$

(117)

with the initial condition F(x,0)=f(x) and we seek non-diverging at infinite times solution F(x,∞)<∞ The solution arises directly from (116), (115) and from (114). Indeed, by noting that Eq. (117) is in fact Eq. (114) with ∂x+α instead of ∂x , we write our solution $F(x,t)=e^{-\alpha x}G(x,t)$ where G(x,t) satisfies Eq. (114) with the initial condition $g(x)=G(x,0)=e^{\alpha x}F(x,0)=e\alpha xf(x)$ and we demand F(x,∞)<∞ and G(x,∞)<∞ respectively. Now the expression (116) for G(x,t) with account for (115), where $g(x)=e^{\alpha x}f(x)$, provide our solution $F(x,t)=e^{-\alpha x}G(x,t)$. Consider the simple example of the initial function $f(x)=e^{-\alpha x}x^n$, which illustrates the above-sketched technique. It returns $g(x)=x^n$ and we easily obtain upon the integration over dudu and dξdξ the function

$$G(x,t)=x^n\exp\left[-t/2\left(\varepsilon+\sqrt{\varepsilon^2+4(n(n-1+\lambda))-\mu}\right)\right].$$ (118)

It, in turn, yields the desired solution $F(x,t)_{|t\to\infty}<\infty$ of (117) with $f(x)=e^{-\alpha x}x^n$, as follows:

$$F(x,t) = x^n \exp\left[-\alpha x - t/2\left(\varepsilon + \sqrt{\varepsilon^2 + 4(n(n-1+\lambda)) - \mu}\right)\right]. \tag{119}$$

The solution for the particular case ε=0ε=0 reads as follows $F(x,t)|_{\varepsilon=0} = x^n e^{-\alpha x - t\sqrt{n(n-1+\lambda)-\mu/4}}$.

Hyperbolic heat equation solution

Another example of the operator D^(x)D^(x) in the r.h.s of Eq. (106) is given by the ∂2x:

$$\left(\frac{\partial^2}{\partial t^2} + \varepsilon\frac{\partial}{\partial t}\right)F(x,t) = \alpha\frac{\partial^2}{\partial x^2}F(x,t), \quad F(x,0) = f(x), F(x,\infty) < \infty, \tag{120}$$

which is one-dimensional case of the Cattaneo's hyperbolic heat conduction equation (Cattaneo 1958)

$$(\tau\partial_t^2 + \partial_t)T = k_T\nabla^2 T, \tag{121}$$

where τ=1/ε is an intrinsic thermal property of the media, characterizing the time needed for the initiation of a heat flow after a temperature gradient appears at the boundary of the domain, and *kT*=α/ε denotes heat diffusivity. The time τ=1/ε is often related to the speed of the second sound *C* in media

$(\tau = k_T/C^2)$; $\sqrt{k_T/\tau} = C$ represents a velocity like quantity, associated with the speed of the heat wave in the medium, which characterizes the thermal wave propagation the same way as the diffusion behaviour is characterized by the diffusivity. Equation (121) is the simplest model of the second sound phenomenon observed first in liquid Helium (Peshkov 1944) and then also in solid crystals (Ackerman and Overton 1969). To solve it we have to compute the result of the action of the operator $\hat{S}$: $e^{-4\alpha\xi\partial_x^2}f(x)$. The fading at infinite time solution for the initial function *f*(x) follows directly from (109):

$$F(x,t) = e^{-\frac{\varepsilon t}{2}} \frac{t}{4\sqrt{\pi}} \int_0^\infty \frac{d\xi}{\xi\sqrt{\xi}} e^{-\frac{t^2}{16\xi} - \varepsilon^2 \xi} \hat{S}f(x), \quad \hat{S}f(x) = e^{-4\alpha\xi\partial_x^2} f(x).$$

(122)

The action of the operator S^S^ can be accomplished with the help of the identity (16):

$$F(x,t) = e^{-\frac{\varepsilon t}{2}} \frac{t}{4\sqrt{\pi}} \int_0^\infty \frac{du}{u\sqrt{u}} e^{-\frac{t^2}{16u} - \varepsilon^2 u} \frac{1}{\sqrt{\pi}} \int_{-\infty}^\infty e^{-v^2} f\left(x + 4iv\sqrt{u\alpha}\right) dv,$$

(123)

Let us consider the initial monomial *f*(x)=x^n, for which we obtain the Hermite polynomials upon the action of the operator S^, as follows from the operational definition (39) with account for (41). Thus, the solution of Eq. (120) takes the following explicit form:

$$F(x,t)|_{F(x,0)=x^n} = e^{-t\varepsilon/2}\frac{t(-i)^n(-4\alpha)^{n/2}}{4\sqrt{\pi}}\int_0^\infty u^{\frac{n-3}{2}}H_m(x/4\sqrt{\alpha u})\exp\left[-\frac{t^2}{16u}-u\varepsilon^2\right]$$

$$= e^{-\frac{t}{2\tau}}\sqrt{\frac{t}{\pi\tau}}n!\sum_{r=0}^{[n/2]}\frac{(x)^{n-2r}}{(n-r)!r!}(-tk_T)^r K_{\frac{1}{2}-r}\left(\frac{t}{2\tau}\right).$$

(124)

Let us now consider the initial function $F(x,0) = e^{\gamma x}x^n$. We make use of the operational identity (52) to obtain the following integral form of the solution:

$$F(x,t)|_{F(x,0)=x^n e^{\gamma x}} = \frac{te^{-\frac{t\varepsilon}{2}+\gamma x}}{4\sqrt{\pi}}\int_0^\infty\frac{H_n(x-8\gamma\alpha u,-4\alpha u)}{u\sqrt{u}}\exp\left[-\frac{t^2}{16u}-u\left(\varepsilon^2+4\alpha\gamma^2\right)\right]du.$$

(125)

For brevity we omit here the result of the above integration, which is rather cumbersome. In the particular case of given *n* and *γ*, for the example for n=−γ=1n=−γ=1, we obtain

$$F(x,t)|_{f(x)=xe^{-x}} = e^{-x-\frac{t\varepsilon}{2}-\frac{t}{2}\sqrt{4\alpha+\varepsilon^2}}\left(x+\frac{2t\alpha}{\sqrt{4\alpha+\varepsilon^2}}\right) = e^{-x-\frac{t}{2\tau}(1+\sqrt{1+4\tau k_T})}\left(x+\frac{2tk_T}{(1+4\tau k_T)^{1/2}}\right)$$

(126)

The example of $F(x,t)|_{f(0,t)=x^{10}e^{-3x}}$ for $n=10,\ \gamma=-3$ is more demonstrative for graphical presentation. We omit here the exact formula for this solution for brevity and present its behaviour for α=ε=1, i.e. kT=1,τ=1 , in Fig. 1. Observe fading wave propagation.

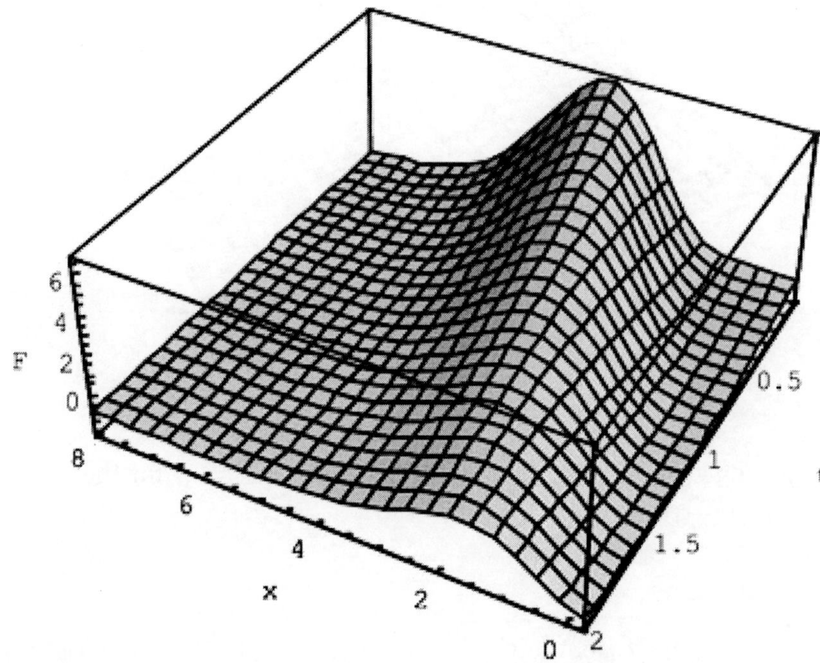

Figure 1. Solution (126) of the Eq. (121) for kT=1,τ=1kT=1,τ=1 for the initial function F(x,0)=e−3xx10F(x,0)=e−3xx10

By choosing α=ε=10 in (120), i.e. kT=1,τ=0.1, we reduce the effect of the second time derivative in the equation, maintaining the heat conductivity unchanged. In this case the fading of the solution happens earlier, as seen in Fig. 2. The diffusive character of the heat conduction in this case prevails over the wave-like propagation process.

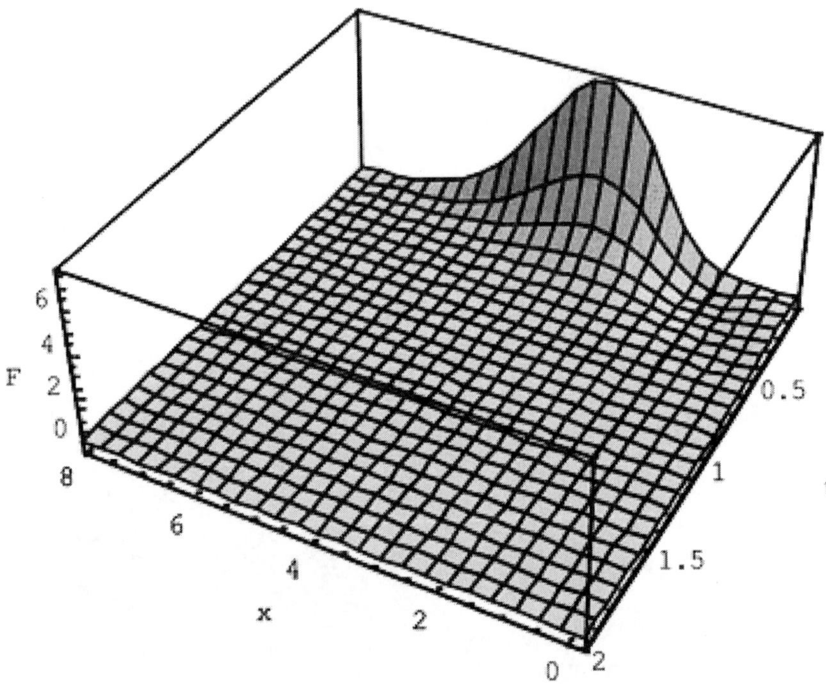

Figure 2. Solution (126) of the Eq. (121) for kT=1,τ=0.1 for the initial function $F(x,0)=e^{-3x}x^{10}$

Contrary to this case, the choice of $\alpha=\varepsilon=0.1$ in (120), i.e. kT=1,τ=10, underlines the wave-like propagation of the initial function in Fig. 3.

Non-Fourier diffusive, wave-like heat propagation in Cattaneo's model (120), (121) had some success in the description of second sound. However, it did not agree with the experimental observations and it was superseded by other, more adequate heat propagation models, which included additional terms in the hyperbolic equation to describe the whole complex of phenomena. We will obtain analytical solutions for them in forthcoming publications.

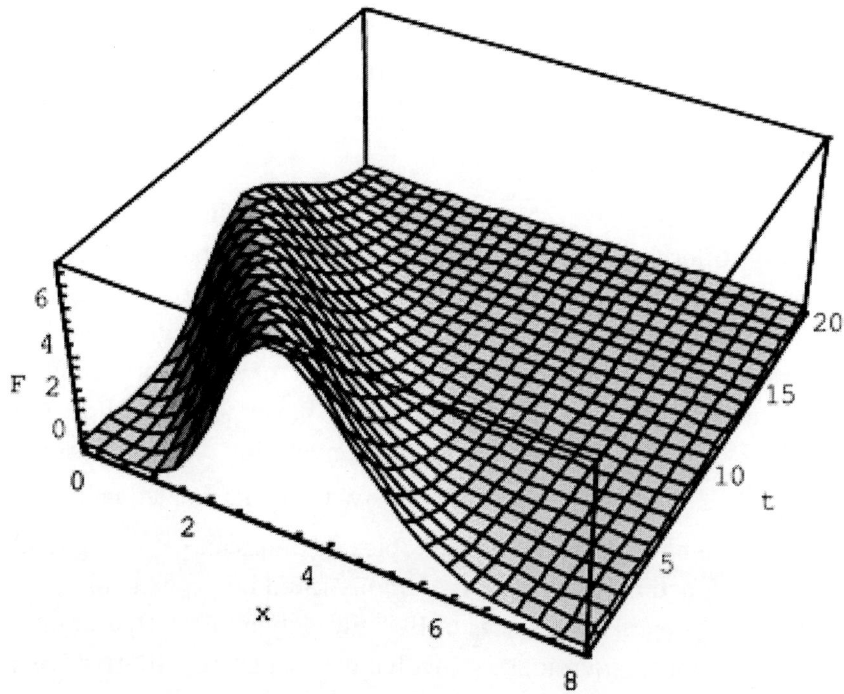

Figure 3. Solution (126) of the Eq. (121) for kT=1,τ=10 for the initial function F(x,0)=e−3xx10

8. RESULTS AND CONCLUSIONS

We advocated operational method for solution of linear differential equations. We use inverse differential operators; it allows direct and straightforward computation of solutions in the framework of operational calculus. The obtained solutions contain consequent action of operators of heat conduction and shift, involving common and Laguerre derivative with exponential and power factors. Their action can be expressed via Gauss type integrals and shift of arguments. Using operational definitions of Hermite and Laguerre orthogonal polynomial families, we executed direct operational transforms over them, consisting in shift and factorization. Combined where necessary with integral transforms, it yielded solutions of relatively complicated linear differential equations of several types.

In particular, we have obtained explicit exact solutions for some ordinary differential equations of non-integer dimension, involving shifted derivatives. The particular solution of equation ψ−1(D)=(β2−(D+α)2)−νf(x) for any Real ν is given by the integral of the weighted consequent action of operators of heat propagation S^S^ and translation Θ^Θ^ on the function f(x)f(x). We wrote it as a convolution transform $\phi(x,\tau)=G(x,\tau)\cdot f(\eta)$ or $\phi=\int_{-\infty}^{\infty}G(x-\eta)f(\eta)$, with the kernel, equal to the Gauss frequency function. The examples of Gaussian $f(x)=e^{-x^2}$ and monomial functions f(x)=xk were demonstrated by explicit solutions in terms of integrals and series of Hermite polynomials. Operational solutions for equations, involving Laguerre derivatives: $(x\partial_x^2+(\alpha+1)\partial_x)^{\nu}F(x)=f(x)$ were obtained. Examples of the exponential f(x)=exp(−γx)f(x)=exp(−γx), of the monomial and of Bessel–Wright functions were demonstrated. By using operational technique we immediately write their explicit solutions; they involve integrals and generalized Laguerre polynomials.

We obtained solutions for several types of partial differential equations. In particular, extended Black–Scholes equation was solved. Moreover, generalized form of Black–Scholes type equation with Laguerre derivative $_LD_x=\partial_x x\partial_x$ was solved operationally. The example of a monomial initial function yields the explicit solution with series of gamma function and hypergeometric function. For initial Bessel–Wright function the solution of the Black–Scholes equation with Laguerre derivative is given by integrals of Bessel–Tricomi function.

Extended forms of heat diffusion equation were solved. Their operational solution readily yields explicit forms upon consequent action of operators of shift and heat diffusion on the initial function. Examples of initial functions $g(x) = x^k e^{\delta x}$ and $f(x,y) = x^m y^n$ produce Hermite polynomials and their generalized forms with four variables and two indices: $H_{m,n}(x, t\alpha, y, t\gamma|\beta)$.). Two-dimensional heat diffusion type equation with the linear terms was solved. Its operational solution consists in the action of the generalized two-dimensional analogue of heat diffusion operator and respective coordinate shifts with a phase factor.

Operational solutions of a number of hyperbolic equations with regular and Laguerre coordinate derivatives were obtained. For the second order of time hyperbolic equations with Laguerre derivatives of the 1st and 2nd order we obtained explicit solutions in terms of elementary functions, Gegenbauer polynomials and gamma function for the initial monomial f(x)=xnf(x)=xn and for the exponential f(x)=xme−αx Hyperbolic heat equation was thoroughly explored with the help of operational method. Explicit solution for f(x)=xme−αx was obtained in elementary functions. The role of various equation terms in the behaviour of the solution was elucidated. Maintaining the heat conductivity unchanged, we underlined the role of the second time derivative. Fading of the wave propagation happens earlier, if we reduce the effect of the second time derivative in the equation, choosing its coefficient small with respect to others: kT=1,τ=0.1. In this case the diffusive character of the heat conduction prevails over the wave-like propagation process. On the contrary, high value of the second time derivative term in (120): kT=1,τ=10 underlines the wave-like propagation with very little fading as follows from the comparison of Figs. 1, 2 and 3.

Thus, operational approach allowed for easy and straightforward solution of differential equations and relevant physical problems, such as modified Fourier heat diffusion in three dimensions, Cattaneo heat propagation, Laguerre type diffusion, evolution of a system, obeying Black–Scholes type equations, common in financial studies. Operational method has obvious advantages, respectively to other methods: it is universal, applies to ordinary and partial linear DE and non-integer DE, the solutions are obtained readily; they are light computationally and have transparent meaning. The effect of each term in the initial equation on the solution is distinguished. The validity of the obtained solutions was verified by direct substitution in the solved equations. The solutions in form of integrals contain consequent action of operators of heat conduction and shift with exponential and power factors. The considered examples of solutions of the hyperbolic and Fourier heat equations with common and Laguerre derivatives for given initial functions contain integrals and series of Hermite polynomials; explicit solutions, such as (96), (97), (105), (113), (119), (126) do not possess critical or singular coordinate points. For the second order of time differential Eqs. (106) with ordinary and Laguerre derivatives we obtained strictly bounded solutions. Rigorous investigation of stability of all of the obtained solutions for different initial functions, including fractional differential equations and employing Lyapunov methods, will constitute a stand-alone study in a forthcoming dedicated publication.

In conclusion we would like to note, that our results, being exact, can represent a benchmark for numerical solutions. These latter can, perhaps, cover more extensions, but should reduce to our results in the limiting cases. Application of our study for more complicated equations, describing non-Fourier heat propagation by ballistic heat transfer and other equations, modelling physical processes, will be also made in forthcoming publications.

COMPETING INTERESTS

The authors declare that they have no competing interests.

REFERENCES

1. Ackerman CC, Overton WC (1969) Second sound in solid helium-3. Phys Rev Lett 22(15):764

2. Akinlar MA, Kurulay M (2013) A novel method for analytical solutions of fractional partial differential equations. Math Probl Eng.

3. Ames WF (2014) Numerical methods for partial differential equations. Academic press, Toronto

4. Appèl A, Kampé de Fériet J (1926) Fonctions Hypergéométriques et Hypersphériques; Polynômes d'Hermite. Gauthier-Villars, Paris

5. Benhammouda B, Vazquez-Leal H (2014) Analytical solutions for systems of partial differential–algebraic equations. SpringerPlus.

6. Carnahan B, Luther HA, Wilkes JO (1969) Applied numerical methods. Wiley, New York

7. Caruntu B, Bota C (2014) Analytical approximate solutions for a general class of nonlinear delay differential equations. Sci World J.

8. Cattaneo C (1958) Sur une forme de l'equation de la chaleur eliminant le paradoxe d'une propagation instantanee. Camptes Rendus de l'Acad. Sc. Paris. 247:431–433

9. Dattoli G (2000) Generalized polynomials, operational identities and their applications. J Comput Appl Math 118:111

10. Dattoli G, Zhukovsky K (2007a) Quark flavour mixing and the exponential form of the Kobayashi–Maskawa matrix. Eur Phys J C 50:817–821

11. Dattoli G, Zhukovsky K (2007b) Quark mixing in the standard model and the space rotations. Eur Phys J C 52(3):591–595

12. Dattoli G, Zhukovsky KV (2008) Neutrino mixing and the exponential form of the Pontecorvo–Maki–Nakagawa–Sakata matrix. Eur Phys J C 55(4):547–552

13. Dattoli G, Srivastava HM, Zhukovsky K (2005) Orthogonality properties of the Hermite and related polynomials. J Comput Appl Math 182:165

14. Dattoli G, Srivastava HM, Zhukovsky K (2006) A new family of integral transforms and their applications. Integral Transform Spec Funct 17(1):31–37

15. Dattoli G, Srivastava HM, Zhukovsky KV (2007) Operational methods and differential equations with applications to initial-value problems. Appl Math Comput 184:979

16. Dattoli G, Mikhailin VV, Zhukovsky K (2008) Undulator radiation in a periodic magnetic field with a constant component. J Appl Phys 104:124507-1–124507-8

17. Dattoli G, Mikhailin VV, Zhukovsky KV (2009) Influence of a constant magnetic field on the radiation of a planar undulator. Mosc Univ Phys Bull 64(5):507–512

18. Demiray ST, Bulut H, Belgacem FB (2015) Sumudu transform method for analytical solutions of fractional type ordinary differential equations. Math Probl Eng.

19. Erdélyi A, Magnus W, Oberhettinger F, Tricomi FG (1953) Higher transcendental functions, vol II. McGraw-Hill Book Company, New York

20. Filobello-Nino U, Vazquez-Leal H, Benhammouda B et al (2015) Analytical solutions for systems of singular partial differential–algebraic equations. Discrete Dyn in Nat Soc

21. Ghia U, Ghia KN, Shin CT (1982) High-Re solutions for incompressible flow using the Navier–Stokes equations and a multigrid method. J Comput Phys 48(3):387–411

22. Gould HW, Hopper AT (1962) Operational formulas connected with two generalizations of Hermite polynomials. Duke Math J 29:51–63

23. Haimo DT, Markett C (1992a) A representation theory for solutions of a higher-order heat equation. I. J Math Anal Appl 168:89

24. Haimo DT, Markett C (1992b) A representation theory for solutions of a higher-order heat equation. II. J Math Anal Appl 168:289

25. Hesam S, Nazemi AR, Haghbin A (2012) Analytical solution for the Fokker–Planck equation by differential transform method. Sci Iran 19(4):1140–1145.

26. Johnson C (2012) Numerical solution of partial differential equations by the finite element method. Courier Corporation, New York

27. Peshkov V (1944) Second sound in Helium II. J. Phys. (Moscow) 8:381

28. Smith GD (1985) Numerical solution of partial differential equations: finite difference methods. Oxford University Press, Oxford

29. Soltanian F, Dehghan M, Karbassi SM (2013) A semi-analytical technique for the solution of differential–algebraic equations and applications in flow of an incompressible viscous fluid. Int J Num Methods Heat Fluid Flow 23(5):818–843.

30. Srivastava HM, Manocha HL (1984) A treatise on generating functions. Wiley, New York

31. Vitanov NK, Dimitrova ZI, Vitanov KN (2015) Modified method of simplest equation for obtaining exact analytical solutions of nonlinear partial differential equations: further development of the methodology with applications. Appl Math Comput 269:363–378

32. Von Rosenberg DU (1969) Methods for the numerical solution of partial differential equations, vol 16. Society of Petroleum Engineers, Richardson

33. Watson GN (1944) A treatise on the theory of Bessel functions, 2nd edn. Cambridge University Press, Cambridge

34. Wolf KB (1979) Integral transforms in science and engineering. Plenum Press, New York

35. Zhukovsky K (2014a) Analytical account for a planar undulator performance in a constant magnetic field. J Electromagn Waves Appl 28(N15):1869–188

36. Zhukovsky K (2014b) Solution of some types of differential equations: operational calculus and inverse differential operators. Sci World J.

37. Zhukovsky KV (2015a) Harmonic radiation in a double-frequency undulator with account for broadening. Mosc Univ Phys Bull 70(4):232–239.

38. Zhukovsky K (2015b) High harmonic generation in undulators for FEL. Nucl Instrum Methods B 369:9–14.

39. Zhukovsky KV (2015c) Harmonic generation by ultrarelativistic electrons in a planar undulator and the emission-line broadening. J Electromagn Waves Appl 29(1):132–142.

40. Zhukovsky K (2015d) High harmonic generation in the undulators for free electron lasers. Opt Commun 353:35–41

41. Zhukovsky KV (2015e) A method of inverse differential operators using ortogonal polynomials and special functions for solving some types of differential equations and physical problems. Mosc Univ Phys Bull 70(2):93–100.

42. Zhukovsky KV (2016) Exact solution of Guyer–Krumhansl type heat equation by operational method Int. J. Heat Mass Transfer 96:132–144.

43. Zhukovsky KV, Dattoli G (2011) Evolution of non-spreading Airy wavepackets in time dependent linear potentials. Appl Math Comput 217:7966–797

CHAPTER 7

Homoclinic and heteroclinic solutions for a class of second-order non-autonomous ordinary differential equations: multiplicity results for stepwise potentials

ABSTRACT

We prove some multiplicity results for a class of one-dimensional nonlinear Schrödinger-type equations of the form

$$x'' + 2kx - 2g(t)x^3 = 0,$$

where k>0 and the weight g(t) is a positive stepwise function. Instead of the cubic term, more general nonlinearities can be considered as well.

KEYWORDS

homoclinic solutions heteroclinic solutions multiplicity nonlinear Schrödinger equation stepwise potential topological methods

DEDICATION

Dedicated to Professor Jean Mawhin

1 INTRODUCTION

This paper deals with the study of homoclinic and heteroclinic solutions for a class of nonlinear second-order differential equations of the form

$$x''+\sigma x-a(t)h(x)=0, \quad (1.1)$$

where $\sigma>0$ is a fixed (positive) coefficient and $a(\cdot):\mathbb{R}\to\mathbb{R}^+_0:=]0,+\infty)$ is a bounded weight function. For the nonlinear term, which we split as

$$h(x)=xf(x),$$

we suppose that

(∗) $f:\mathbb{R}\to\mathbb{R}^+:=[0,+\infty)$ *is a locally Lipschitz function which is even, strictly increasing on* $[0,+\infty)$ *such that* $f(0)=0$ *and* $f(+\infty)=+\infty$.

Examples of functions f which are suitable for our considerations are, for instance, $f(s)=s^2$ as in [1], or $f(s)=s^2+bs^4$ $(b>0)$ as in [2]. In both these cases, the function f is even. In Section 3 we briefly describe how this restriction could be avoided.

Equations of the form (1.1) naturally arise in the search of particular solutions for some classes of nonlinear Schrödinger equations (NLSE) with inhomogeneous nonlinearities. Typical examples are related to the NLSE

$$i\hbar\frac{\partial\Psi}{\partial t} = -\frac{\hbar^2}{2m}\Delta\Psi + V_1(x)\Psi + V_2(x)|\Psi|^{p-1}\Psi, \quad x\in\mathbb{R}^N, \tag{1.2}$$

where $\hbar$ is the Planck constant, m is the particle's mass, i is the imaginary unit, Δ is the Laplace operator and $p>1$. In literature, solutions of the form

$$\Psi(t,x) = u(x)\exp\left(i\lambda\hbar^{-1}t\right),$$

where $\lambda \in R$ and $u(x)$ is a real-valued function, are called *stationary waves*. Their search leads to the equation

$$-\frac{\hbar^2}{2m}\Delta u + \lambda u + V_1(x)u - V_2(x)|u|^{p-1}u = 0, \quad x\in\mathbb{R}^N \tag{1.3}$$

(see [3]). Usually, the Planck constant and the mass are omitted in (1.3) after rescaling.

In many significant models of NLSE, one-dimensional waves are considered. They are studied, for example, in nonlinear optics, in the theory of ocean rogue waves and for Bose-Einstein condensates (just to mention a few cases). For instance, in [1], Belmonte-Beitia and Torres analyzed the one-dimensional equation

$$-\frac{1}{2}u'' + \lambda u + g(x)u^3 = 0, \quad x \in \mathbb{R}, \tag{1.4}$$

which is a particular case of (1.3) with N=1, p=3$p = 3$ and V1=0. In equation (1.4) the nonlinearity is called *inhomogeneous* due to the presence of a non-constant weight $V_2(x)=g(x)$, which in [1] is assumed to be positive. The sign condition on g(x) implies that nontrivial bounded solutions of (1.4) can exist only for $\lambda<0$ (see [[1], Theorem 1]). For this reason, we prefer to set $\lambda=-k$ with k>0.

In order to study equation (1.4) or its variants, we are going to follow a dynamical system approach, hence we choose to treat the independent variable (which in the applications has a spatial connotation) as a time variable, via the substitution $x \leftrightarrow t$. Similarly, for the dependent variable, we make the substitution $u \leftrightarrow x$. In this way, equation (1.4) reads as

$$x'' + 2kx - 2g(t)x^3 = 0, \tag{1.5}$$

which belongs to the same class of (1.1).

In spite of the apparent simplicity of equation (1.5), a throughout study of its solutions may be a rather difficult task for a general nonconstant weight function g(t). Looking for homoclinic and heteroclinic solutions of (1.5), it will be natural to focus on the behavior of $g(t)$ at $\pm\infty$. In similar situations, various authors have confined their study to the case in which $g(t)$ is asymptotically constant [4–6], or eventually constant [7, 8]. These assumptions are also justified by the analysis of some physical underlying models, in which a layered structure is present. With this respect, see the introduction in [8], where different eventually autonomous cases are listed for related NLSEs arising in nonlinear optics. Examples in which the nonlinear term presents a piecewise constant weight function have been studied also in biological and chemical models. In particular, these situations occur in the theory of wave propagation for reaction-diffusion systems; see, for instance, [9–11]. In the context of equation (1.5), examples in which $g(t)$ is a piecewise constant function have been considered as well (see [12]).

The aim of the present paper is to provide multiplicity results regarding homoclinic and heteroclinic solutions for equation (1.1) under particular assumptions on the weight function $a(t)$. Actually, we suppose that $a(t)$ is an eventually constant piecewise function with only two steps.

Our approach combines phase-plane analysis with time-mapping estimates. As in [5, 7, 8, 10, 11, 13], the solutions are obtained by connecting the unstable and stable manifolds of the equilibrium points of the asymptotically autonomous equations. Such connections are performed by means of orbit paths of an intermediate equation, which represents the behavior of the system during a suitable interval of transition $[t-,t+]$ between the asymptotic states. Multiple connecting solutions arise when such interval length is sufficiently large. Lower estimates for $t+-t-$ will be provided in terms of time mappings, which can be expressed by Abelian integrals. This kind of approach is also reminiscent of some topological methods for the study of Sturm-Liouville boundary value problems. Indeed, a solution that satisfies the Sturm-Liouville boundary conditions can be interpreted as a trajectory in the phase-plane that connects two lines (see [14–17]). Generalized Sturm-

Liouville solutions, which connect the graphs of two functions or given planar continua, have been considered as well (see [18–20]). Often these problems can be settled in the framework of the theory of ODEs with *nonlinear boundary conditions* (see [21, 22]).

In order to make our approach more transparent, we are going to perform our analysis for equation (1.5). This choice is motivated by the sake of avoiding unnecessary technicalities. Our arguments can be modified in a straightforward manner in the case of more general equation (1.1), with f satisfying $(*)$ (see Section 3). Homoclinic and heteroclinic orbits can be interpreted as solutions for some boundary value problems on unbounded intervals. In the last section we also outline possible applications of our approach to boundary value problems on a compact interval (like the Sturm-Liouville one).

Besides his manifold achievements in different areas of mathematics, Professor Jean Mawhin is one of the pioneers in the study of topological methods for nonlinear boundary value problems. It is a pleasure and an honor to have the possibility to dedicate our work to his important contributions in this area.

2 HOMOCLINIC AND HETEROCLINIC SOLUTIONS: MULTIPLICITY RESULTS

2.1 General setting

We consider the second-order nonlinear differential equation

$$x'' + 2kx - 2g(t)x^3 = 0 \tag{2.1}$$

and its equivalent system in the phase-plane

$$\begin{cases} x' = y, \\ y' = -2kx + 2g(t)x^3, \end{cases} \tag{2.2}$$

where k>0$k > 0$ is a fixed coefficient and g:R→R+0:=]0,+∞) is a bounded measurable function. Solutions of (2.1) are meant in the generalized (*i.e.*, Carathéodory) sense (see [23]). Actually, in our results a step function g(t)$g(t)$ with only two jumps is considered and the solutions are piecewise smooth.

In the particular case of a constant coefficient g(t)≡μ>0$g(t) \equiv \mu > 0$, an elementary analysis of the system

$$\begin{cases} x' = y, \\ y' = -2kx + 2\mu x^3 \end{cases} \tag{2.3}$$

shows that the associated phase portrait presents a center O=(0,0)$O = (0,0)$ and two saddle points P $P_\pm := (\pm(k/\mu)^{\frac{1}{2}}, 0)$.. These saddles are connected by two heteroclinic solutions (one connecting P− to P+ in the upper half-plane and a symmetric one from P+ to P−P_- in the lower half-plane). The heteroclinic orbits are described by

$$y^2 = \mu\left(x^2 - (k/\mu)\right)^2 \quad \text{for } -(k/\mu)^{\frac{1}{2}} < x < (k/\mu)^{\frac{1}{2}}. \tag{2.4}$$

In [1] Belmonte-Beitia and Torres supposed that the weight function g(t)$g(t)$ is *even and T-periodic* and proved the existence of a heteroclinic solution connecting two periodic stationary states. Such a result generalizes to the case of periodic coefficients the situation described above for the autonomous system.

As already observed in the introduction, equation (2.1) has been already studied by various authors for its relevance in many applicative models. The aim of the present paper is to provide multiplicity results for heteroclinic and also homoclinic solutions of equation (2.1). A possible way to obtain this goal is to assume that the weight function g(t) has a different behavior at infinity and in some intermediate time interval. In this setting, a first step consists of analyzing the situation in which g(t) is a stepwise function assuming only two values. We believe that the study of such a simplified case may lead to more general considerations, in which, for instance, g(t) is eventually periodic. In this way, multiplicity results for Belmonte-Beitia and Torres's model can be obtained (see Section 3 for a brief overview concerning possible applications of our approach). Equation (2.1) with a *periodic stepwise* coefficient g(t)$g(t)$ has recently been studied in [24] in the context of chaotic-like dynamics.

As a tool, in our paper, we use the Poincaré map associated to system (2.2). Given a fixed time interval [t0,t1] and a point z∈R², we indicate by ζ(·,t0,z) the solution ζ(t)=(x(t),y(t)) of (2.2) satisfying the initial condition ζ(t0)=z. The Poincaré map on [t0,t1] is defined as

$$\Phi = \Phi_{t_0}^{t_1} : z \mapsto \zeta(t_1, t_0, z).$$

From the fundamental theory of differential equations it follows that the domain DΦ of Φ is an open subset of the plane and Φ is an orientation preserving homeomorphism of DΦ onto its image Φ(DΦ).

2.2 Analysis of the equation

Let us consider system (2.2) for the stepwise weight function g(t)$g(t)$, defined as

$$g(t) = \begin{cases} \mu_1 & \text{for } t \leq t_- \text{ or } t \geq t_+, \\ \mu_0 & \text{for } t_- < t < t_+, \end{cases} \tag{2.5}$$

where

t−<t+and0<μ0<μ1.

System (2.2) can be seen as the superposition of the autonomous systems

$$\begin{cases} x' = y, \\ y' = -2kx + 2\mu_1 x^3 \end{cases} \tag{2.6}$$

and

$$\begin{cases} x' = y, \\ y' = -2kx + 2\mu_0 x^3. \end{cases} \tag{2.7}$$

The first one describes the asymptotic behavior of the equation. We are going to use the orbits of the second system for connecting unstable/stable manifolds of (2.6) during the time interval [t−,t+]. Figure 1 shows the superposition of the phase portraits of the two systems.

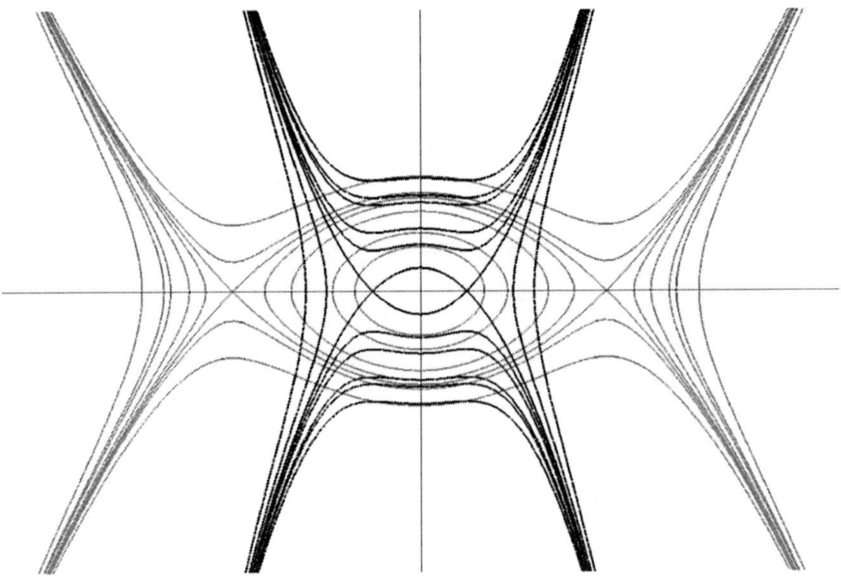

Figure 1. In The Present Figure We Have Considered The Superposition Of The Phase Portraits Of Systems **(2.6)** (In Darker Color) And **(2.7)** (In Lighter Color) With**k**=1,μ1=4,μ0=1/4. For graphical reasons, a slightly different *x*- and *y*-scaling has been used.

First of all, we briefly analyze the structure of system (2.6). Its equilibrium points are the origin (which is a center) and two saddle points P±=(±ξ1,0), where

$$\xi_1 := (k/\mu_1)^{\frac{1}{2}}.$$

The stable and unstable manifolds of P± are illustrated in Figure 2. The sets I(P−)∪L(P+,P−) and O(P−)∪L(P−,P+) are, respectively, the stable and the unstable manifolds of P−. Symmetrically, I(P+)∪L(P−,P+) and O(P+)∪L(P+,P−) are the stable and the unstable manifolds of P+.

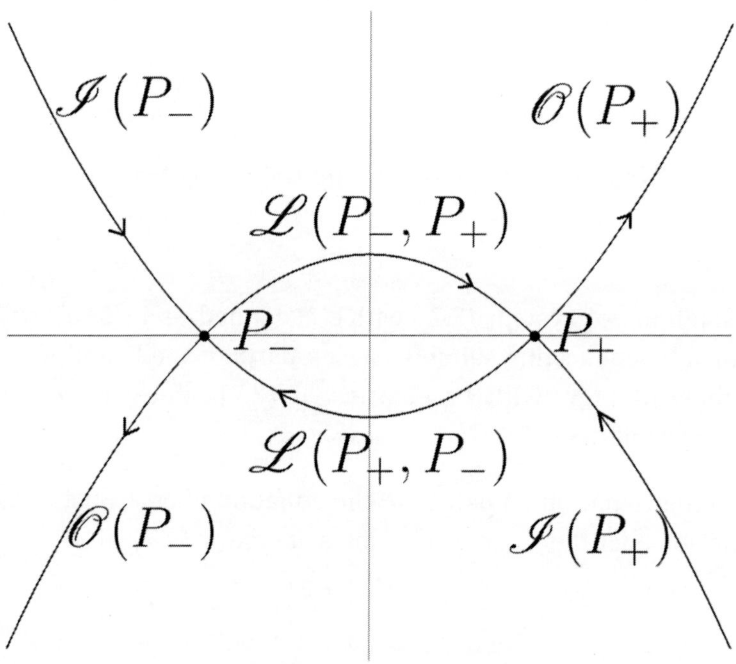

Figure 2. The Stable And Unstable Manifolds Of**p−**P_And **p+** The arrows show the direction of the flow along the orbits.

In the case of the cubic nonlinearity, we find

$$\mathcal{I}(P_-) = \{(x,y) : y = |\mu_1 x^2 - k| / \sqrt{\mu_1}, x < -(k/\mu_1)^{\frac{1}{2}}\},$$

$$\mathcal{O}(P_-) = \{(x,y) : y = -|\mu_1 x^2 - k| / \sqrt{\mu_1}, x < -(k/\mu_1)^{\frac{1}{2}}\},$$

$$\mathcal{I}(P_+) = \{(x,y) : y = -|\mu_1 x^2 - k| / \sqrt{\mu_1}, x > (k/\mu_1)^{\frac{1}{2}}\},$$

$$\mathcal{O}(P_+) = \{(x,y) : y = |\mu_1 x^2 - k| / \sqrt{\mu_1}, x > (k/\mu_1)^{\frac{1}{2}}\},$$

$$\mathcal{L}(P_-, P_+) = \{(x,y) : y = |\mu_1 x^2 - k| / \sqrt{\mu_1}, -(k/\mu_1)^{\frac{1}{2}} < x < (k/\mu_1)^{\frac{1}{2}}\},$$

$$\mathcal{L}(P_+, P_-) = \{(x,y) : y = -|\mu_1 x^2 - k| / \sqrt{\mu_1}, -(k/\mu_1)^{\frac{1}{2}} < x < (k/\mu_1)^{\frac{1}{2}}\}.$$

The solutions of system (2.2) with g(t) defined as in (2.5) will be obtained by connecting suitably chosen parts of stable and unstable manifolds of (2.6) with trajectories of (2.7). More in detail, we proceed as follows.

Due to the Hamiltonian nature of the equation under consideration, we can consider the 'energy' level lines of system (2.7), given by

$$E(x,y) := y^2 + 2kx^2 - \mu_0 x^4 = c. \tag{2.8}$$

We also set

$$\Gamma^c := \{(x,y) : -\xi_0 \le x \le \xi_0, E(x,y) = c\},$$

Where

$$\xi_0 := (k/\mu_0)^{\frac{1}{2}}.$$

The curve ΓcΓ^c is the part of the line at level c∈R$c \in \mathbb{R}$ in the strip [−ξ0,ξ0]×R. For every *c* such that

$$0 < c < \ell_0 := \frac{k^2}{\mu_0},$$

the set ΓcΓ^c is a periodic orbit of (2.7) run in the clockwise sense, which intersects the *y*-axis at the points $(0, \pm\sqrt{c})$The (minimal) period τc$\tau_c$of ΓcΓ^c can be expressed by the following time-mapping formula:

$$\tau_c = 2\int_{s_0}^{s_1} \frac{dx}{\sqrt{c - 2F(x)}}, \tag{2.9}$$

where F(x) is the potential associated to the equation and s0s_0, s1s_1 are such that

$$F(s_0) = F(s_1) = c/2 \quad \text{with} \ -\xi_0 < s_0 < 0 < s_1 < \xi_0.$$

In our case,

$$F(x) = kx^2 - \frac{\mu_0}{2}x^4$$

and by symmetry $s_0=-s_1$. More in detail, we can express the period τ_c by means of an elliptic integral in the following way:

$$\tau_c = \frac{4}{\mu_0}\int_0^{s_1} \frac{dx}{\sqrt{(a^2-x^2)(b^2-x^2)}}, \tag{2.10}$$

for $a := s_1,\ b := \left(\frac{2k^2}{\mu_0^2} - s_1^2\right)^{1/2}$ and $0 < s_1 < \xi_0,\ 0 < c = 2ks_1^2 - \mu_0 s_1^4 < \ell_0$.
Notice that $0<a<b$ in (2.10). Therefore

$$\tau_c = \frac{4}{b\mu_0}\,\mathrm{sn}^{-1}(1, a/b),$$

where $\mathrm{sn}(\cdot,k)$ is the *Jacobi elliptic sine function of modulus* $k:=a/b$ (see [25, 26]).

The level line Γ_{ℓ_0} contains the saddle points $Q_\pm:=(\pm\xi_0,0)$ of system (2.7) and their heteroclinic connections $L(Q_-,Q_+)$ in the upper half-plane and $L(Q_+,Q_-)$ in the lower half-plane. The energy level

$$\ell_1 := \frac{k^2}{\mu_1}$$

corresponds to the closed curve $\Gamma\ell^1$, which is tangent to L(P−,P+) and L(P+,P−). Moreover, for

$$\ell^* := \frac{k^2}{\mu_1}\left(2 - \frac{\mu_0}{\mu_1}\right),$$

the energy level line $\Gamma\ell*$ contains the saddle points P± of system (2.6).

Using the parameters $\ell 0$, $\ell 1$ and $\ell *$, we define the regions

$$\mathcal{E} := \left\{(x,y) : -\xi_0 \le x \le \xi_0, \ell_1 \le E(x,y) \le \ell^*\right\},$$

$$\mathcal{F} := \left\{(x,y) : -\xi_0 \le x \le \xi_0, \ell^* < E(x,y) < \ell_0\right\}.$$

Both $\mathcal{E}$ and $\mathcal{F}$ are invariant sets for (2.7); they are filled by periodic orbits. In the sequel, the orbits in the region $\mathcal{E}$ will be called *internal*, while those in $\mathcal{F}$ will be called *external*. The choice of these names is made in order to distinguish the trajectories with respect to $\Gamma\ell*$, which contains the saddle points P± of system (2.6). (See Figure 3.)

In order to obtain homoclinic or heteroclinic solutions for system (2.2) with g(t) defined as in (2.5), we connect the unstable manifold of the point A∈{P−,P+} to the stable manifold of B∈{P−,P+} via an orbit path of Γc for c∈[ℓ1,ℓ0[. Actually, this is the only way to get the desired solutions. If c≤ℓ∗, such connection will follow a trajectory (x(t),y(t)) with −ξ0≤x(t)≤ξ0, so it will lie in the invariant annular region $\mathcal{E}$ bounded by Γ^{ℓ_1} and $\Gamma\ell*$. On the contrary, for c>$\ell*$ the connection lies in the set $\mathcal{F}$ between $\Gamma\ell*$ and $\Gamma\ell0$.

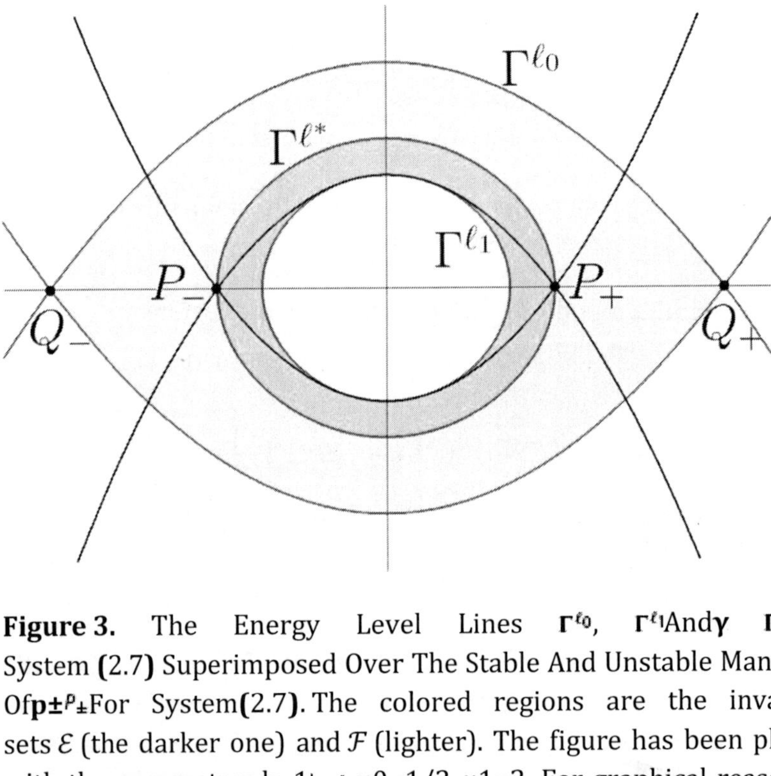

Figure 3. The Energy Level Lines $\boldsymbol{\Gamma^{\ell_0}}$, $\boldsymbol{\Gamma^{\ell_1}}And\boldsymbol{\gamma}$ $\boldsymbol{\Gamma^{\ell^*}}$For System **(**2.7**)** Superimposed Over The Stable And Unstable Manifolds Of**p±**$^{P_\pm}$For System**(**2.7**)**. The colored regions are the invariant sets $\mathcal{E}$ (the darker one) and $\mathcal{F}$ (lighter). The figure has been plotted with the parameters k=1$k=1$, μ0=1/2, μ1=2. For graphical reasons, a slightly different *x*- and *y*-scaling has been used.

To provide more explicit details, we consider the case A=P+. A solution of (2.2), which is homoclinic to P+, can be produced in two ways. One consists in connecting O(P+) to I(P+), using an orbit path of Γc for c∈]$\ell*$,$\ell0$[. Another possibility is given by connecting a point ofL(P+,P−) to a point of L(P−,P+) via an orbit path of Γc for c∈[$\ell1$,$\ell*$[. For c=$\ell*$, we can obtain our solution if and only if t+−t−is an integer multiple of the fundamental period $\tau\ell*$ of the orbit $\Gamma\ell*$. Indeed, such a solution will be constantly equal to P+ for t∉]t−,t+[, and for t∈[t−,t+] it coincides with the periodic solution of (2.7) such that x(t−)=ξ1 and y(t−)=0. This solution makes (t+−t−)/$\tau\ell*$turns around the origin for t∈[t−,t+].

Similar considerations can be developed with respect to heteroclinic solutions. For instance, a heteroclinic orbit from P_+ to P_- can be obtained as follows: by connecting a point of with one of using an orbit path of Γ^c for c∈][, or by a connection of two different points of $\mathcal{L}(P_+,P_-)$ via an orbit path of Γ^c for c∈[ℓ1,ℓ*[. For c=ℓ*, we can obtain our solution if and only if t+−t− is an odd multiple of $\tau\ell*/2$, where $\tau\ell*$ is the period of the orbit $\Gamma\ell*$. Indeed, such a solution will be constantly equal to P+ for $t\in$(−∞,t−] and constantly equal to P−P_- for t∈[t+,+∞). Moreover, it coincides for t∈[t−,t+] with the solution of (2.7) such that x(t−)=ξ1 and y(t−)=0. This solution makes 2(t+−t−)/$\tau\ell*$ half-turns around the origin for t∈[t−,t+].

Figure 4, although not exhaustive of all the conceivable cases, summarizes several different possibilities. Indeed, a trajectory homoclinic to P+P_+ can be obtained as follows: move by system (2.6) along O(P+) from P+ to the intersection point of O(P+) with a closed curve external to P+P_+, namely δ (for t∈(−∞,t−]). This point is put in evidence with a small black circle. Next, follow δ (by system (2.7)) in order to reach the intersection of δ with I(P+) (for $t \in [t_-, t_+]$). Such an intersection point is indicated by a grey square. Finally, switch to system (2.6) and move toward P+ along I(P+) (for t∈[t+,+∞)). With a similar procedure, we can obtain a heteroclinic trajectory from P+ to P−. In fact, we can move by system (2.6) along O(P+) from P+ to the intersection point of O(P+) with a closed curve external to P+, namely γ (for t∈(−∞,t−]). Next, follow γ (by system (2.7)) to the intersection with I(P−) (for t∈[t−,t+]). Finally, switch to system (2.6) and move toward P− along I(P−) (for t∈[t+,+∞)).

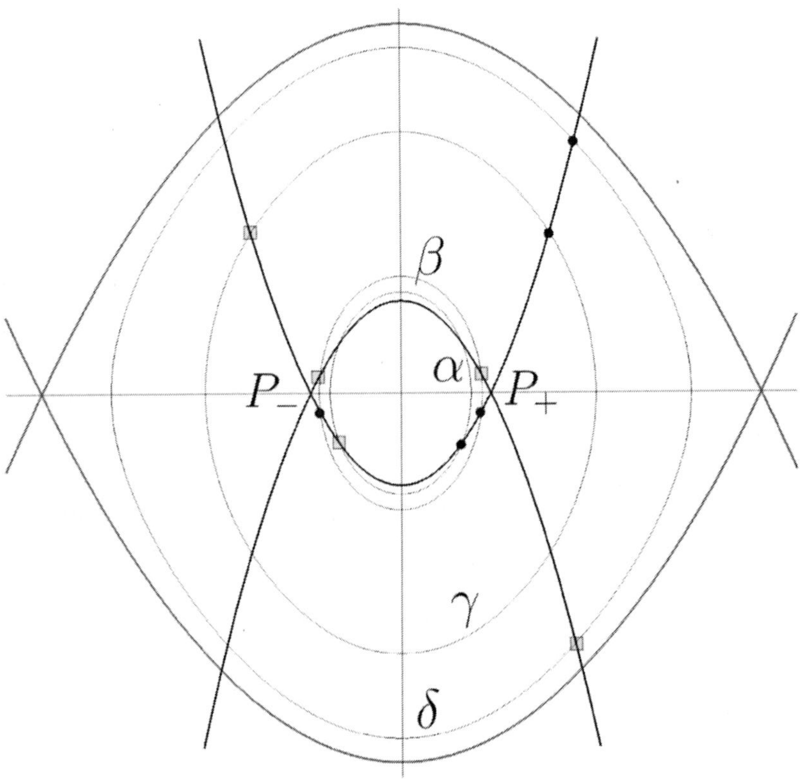

Figure 4. In The Present Figure We Have Considered The Superposition Of **(**2.6**)** And **(**2.7**)** For**k**=1, μ1=4, μ0=1/4**.** The closed curvesα, β, γ, δ represent four different level lines of system (2.7): the curves α and β in $\mathcal{E}$ correspond to a level c∈]ℓ1,ℓ∗[, while γ and δ in$\mathcal{F}$ refer to c∈]ℓ∗,ℓ0[. In order to describe a connection from an unstable manifold to a stable one of system (2.6) via an orbit path of (2.7), we have marked with a black circle possible starting points and with a grey square some available end points on the same level lines.

In general, the connections through the trajectories of system (2.7) will either involve only an arc of the closed curve Γc, or they will require to perform a certain number of winds around the origin, depending on the available time. If the length of the time interval [t−,t+] is small, only the first case is possible. However, if t+−t− is large enough, more choices arise. For instance, looking at the orbit δ in Figure 4, we could make a certain number of loops from the black circle before reaching the grey square on the same orbit. Similar considerations apply to the heteroclinic trajectory described above with reference to γ.

Until now we have described the 'external connections', namely those which lie in the region $\mathcal{F}$. Further possibilities appear if we consider 'internal connections' by means of orbit paths contained in $\mathcal{E}$. For example, if we look at the orbit β in Figure 4, we can obtain trajectories homoclinic to P+ by choosing one of the two black circles as a starting point from L(P+,P−) and one of the two grey squares as an end point on L(P−,P+). Loops along β will be permitted if the time is sufficiently large. The construction of heteroclinic orbits from P+P_+ to P−P_-via the line α follows a similar procedure.

2.3 Study of the Poincaré map

Another possible point of view to describe the previous construction of homoclinic and heteroclinic solutions consists in considering the Poincaré map associated to system (2.7) for the time interval [t−,t+]. Due to the autonomous nature of this system, we have that $\Phi_{t_-}^{t_+} = \Phi_0^{t_+ - t_-}$ denoted simply by Φ when no confusion may occur. Moreover, the region

$$\mathcal{W} := \{(x,y) : -\xi_0 \leq x \leq \xi_0, E(x,y) \leq \ell_0\}$$

is a compact invariant set contained in the domain of Φ. Let us denote by $\mathcal{Y} := \overline{\mathcal{O}(P_+)} \cap \overline{\mathcal{F}}$ the part of the unstable manifold O(P+)between P+P_+ and $\Gamma^{\ell 0}$, including the extreme points. Hence the external connections can be precisely described by looking at the intersection points of Φ(Y) with I(P+) (for the homoclinics) and with I(P−) (for the heteroclinics), as illustrated in Figure 5.

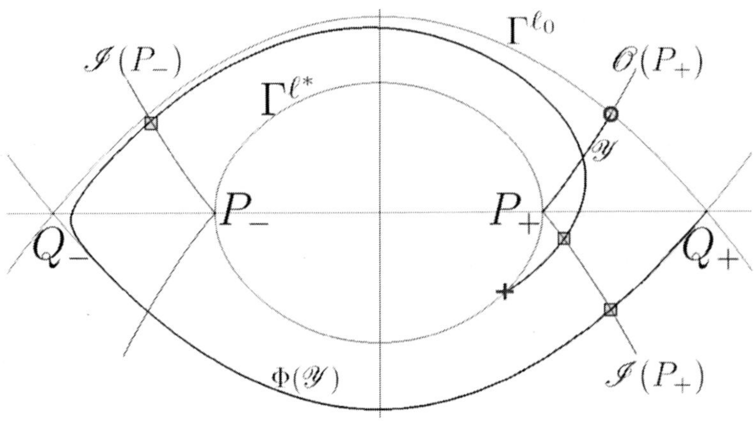

Figure 5. In The Present Figure We Show The Transformation Of $\mathcal{Y}$ Under The Poincaré Map Φ For**t**+−t−=5.5**.** The parameters used arek=1 $k=1$, μ1=2 , μ0=1/2. For graphical reasons a slightly different *x*- and *y*-scaling has been used.

With reference to Figure 5, we observe that the intersection point of O(P+) with $\Gamma^{\ell 0}$ (indicated by a small circle) is moved by Φ along $\Gamma^{\ell 0}$ to a point very close to Q+. On the other hand, during the time interval [t−,t+], the point P+ makes a little more than a complete turn around the origin; the final position Φ(P+) is indicated by a cross. The arc Φ(Y) is a spiral-like curve with one end near Q+ and the other on Γℓ∗. For the time length t+−t−=5.5, considered in

Figure 5, there are precisely two intersections of $\Phi(Y)$ with $I(P+)$, corresponding to external homoclinic solutions. Meanwhile, we have also an external heteroclinic solution due to the intersection of $\Phi(Y)$ with $I(P-)$. All these three intersection points have been indicated by a grey square (following the same convention used in Figure 4).

In order to describe further these solutions, let us consider their energy at the time $t+t_+$ that we denote by k_1, k_2, k_3, with $\ell_* < k_3 < k_2 < k_1 < \ell_0$. Let us define the points H_i as $\{H_i\} = O(P+) \cap \Gamma k_i$ for $i=1,2,3$. The solution $\zeta(t)=(x(t),y(t))$ of (2.2) with $\zeta(t-)=H1$ or $\zeta(t-)=H3$ has the following behavior: $\zeta(t) \to P+$ for $t \to \pm\infty$, with $x(t) > \xi 1$ and convex for $t \notin]t-,t+[$. Moreover, $x(t)$ is increasing on $(-\infty,t-]$ and decreasing on $[t+,+\infty)$. On the interval $[t-,t+]$, if $i=1$, the solution $x(t)$ is concave, while if $i=3 i=3$, the solution has two maxima and one (negative) minimum separated by two simple zeros. The solution of system (2.2) with $\zeta(t-)=H2$ is such that $\zeta(t) \to P+$ for $t \to -\infty$ with $x(t) > \xi 1$ and $\zeta(t) \to P-$ for $t \to +\infty$ with $x(t) < -\xi 1$; $x(t)$ is convex for $t \notin]t-,t+[$. Moreover, the solution has one maximum and one minimum separated by one simple zero.

Even if the time interval length $t+-t-$ is small, we always have at least one intersection of $\Phi(Y)$ with $I(P+)$ and thus a homoclinic solution. On the other hand, if $t+-t- t_+ - t_-$ grows, the spiral curve $\Phi(Y)$ will wind more times around the origin, hence more homoclinic/heteroclinic solutions will appear.

If we look for the internal connections (made by orbit paths lying in the region $\mathcal{E}$), we proceed as follows.

Let us set $\mathcal{L} := \overline{\mathcal{L}(P_+, P_-)}$ and denote by $\mathcal{L}' := \overline{\mathcal{L}(P_-, P_+)}$ its symmetric part with respect to the x-axis. Figure 6 illustrates the involved geometry from the point of view of the Poincaré map. We observe that all the points of $\mathcal{L}$ are contained in $\mathcal{E}$, hence they are periodic points (of different periods). As before, during the time interval [t−,t+], the point P+ moves clockwise around the origin along Γℓ∗; the final position Φ(P+)∈Γℓ∗ is indicated by a cross. Symmetrically, the point P−P_- performs exactly the same angle around the origin, reaching the final position Φ(P−)∈Γℓ∗, indicated by a small circle. All the other points of L(P+,P−) move, under the action of Φ, on the energy level lines ΓcΓ^c with ℓ1≤c<ℓ∗. The arc Φ(L) is a curve contained in $\mathcal{E}$, connecting the two antipodal points Φ(P±)∈Γℓ∗ and leaning on one point of Γℓ1. This tangent point is the image through Φ of the point $R := (0, -\sqrt{\ell_1})$. Such a property of Φ(L) implies that Φ(L)intersects both $\mathcal{L}$ and L′$\mathcal{L}'$ (at different points). These two points have been indicated by a grey square in the figure. Accordingly, we always find at least an internal homoclinic solution and an internal heteroclinic one.

Summarizing the above information, we conclude that, for every time interval [t−,t+], we have at least *two solutions* which are homoclinic to P+ (one external and one internal) and one internal heteroclinic solution from P+ to P−.

For the *external* connections, if the time interval length t+−t− grows, the situation can be summarized as follows. The number of *external*homoclinic/heteroclinic solutions increases, depending on the number of winds around the origin of the curve Φ(Y). Indeed, the end point of such a curve on Γℓ0 cannot go beyond Q+, while the other end point Φ(P+) is free to move on the periodic orbit Γℓ∗. If we denote by η_{ext} and ν_{ext} the number of external homoclinic and heteroclinic solutions respectively, it holds that $0 \le \eta_{ext} - \nu_{ext} \le 1$. From the above discussion, we can conclude that if, for some nonnegative integer *n*, we have t+−t−>nτℓ∗ (where τℓ∗ is the period of Γℓ∗), then ηext≥n+1. An analogous lower bound can be provided for ν_{ext}. For a formal proof, see Theorem 2.1.

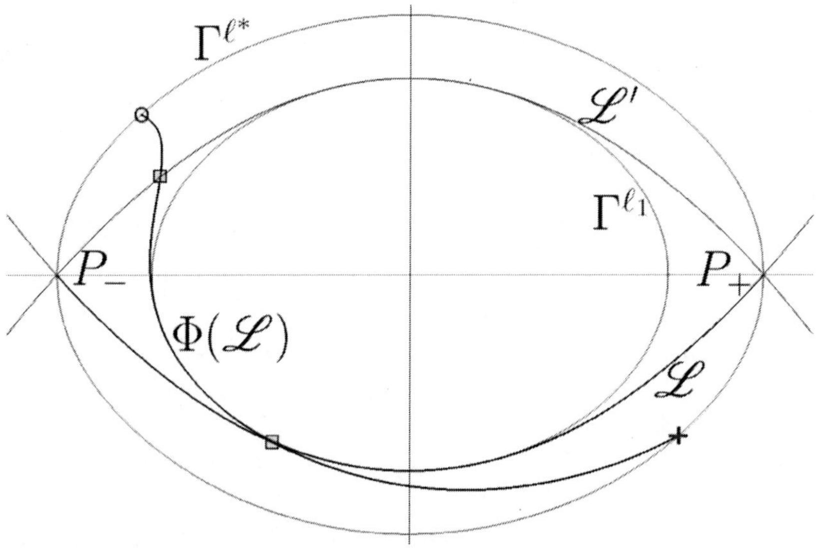

Figure 6. In The Present Figure We Show The Transformation Of$\mathscr{L} := \overline{\mathscr{L}(P_+, P_-)}$Under The Poincaré Map Φ For**t+−T−=5.5**, The Same Time Considered In Figure**5.** The parameters used are k=1$k=1$, μ1=2, μ0=1/2, as before. The point indicated by a cross is precisely the same point Φ(P+) of Figure 5. Notice that for this time interval the point makes a little more than a complete turn around the origin. For graphical reasons, a slightly different *x*- and *y*-scaling has been used.

When t+−t− grows, the situation for the *internal* homoclinic/heteroclinic solutions is more intriguing, as illustrated in Figure 7 and Figure 8. A twist effect depending on the different periods of Γℓ1 and Γℓ∗ produces a double spiral-like curve Φ(L) when t+−t− is sufficiently large.

The fact that Φ(L) looks like a double spiral depends on the different velocities of the points P±$P_\pm$ and $R = (0, -\sqrt{\ell_1})$ (which is the intersection of $\mathcal{L}$ with the negative *y*-axis). In fact, since τℓ1<τℓ∗, the points on $\Gamma\ell 1\Gamma^{\ell_1}$ move faster than those of $\Gamma\ell*$. By construction, $P\pm\in\Gamma\ell*$, while $R\in\Gamma\ell 1 R \in \Gamma^{\ell_1}$. Therefore, when the time gap t+−t− is sufficiently large, the number of turns of Φ(R) around the origin will exceed the number of turns of Φ(P±). Accordingly, the image through Φ of the right part of $\mathcal{L}$ (the sub-arc of $\mathcal{L}$ connecting P+P_+ to *R*) is a spiral that winds a certain number of times in the clockwise sense around the origin. Indeed, it connects the slower point Φ(P+) to the faster one Φ(R)$\Phi(R)$ inside the region $\mathcal{E}$, as illustrated in Figure 9. The number of half-turns of this spiral depends on a 'rotational gap' that, for our purposes, it will be convenient to define as

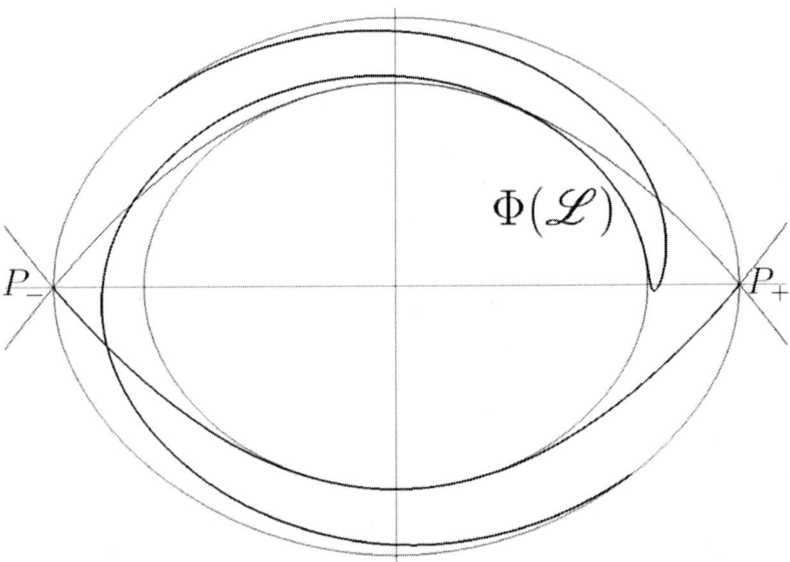

Figure 7. In The Present Figure We Show The Transformation Of $\mathcal{L}$ Under The Poincaré Map Φ For**t**+−t−=50. The parameters used are k=1, μ1=2, μ0=1/2, as before. For graphical reasons, a slightly different *x*- and *y*-scaling has been used.

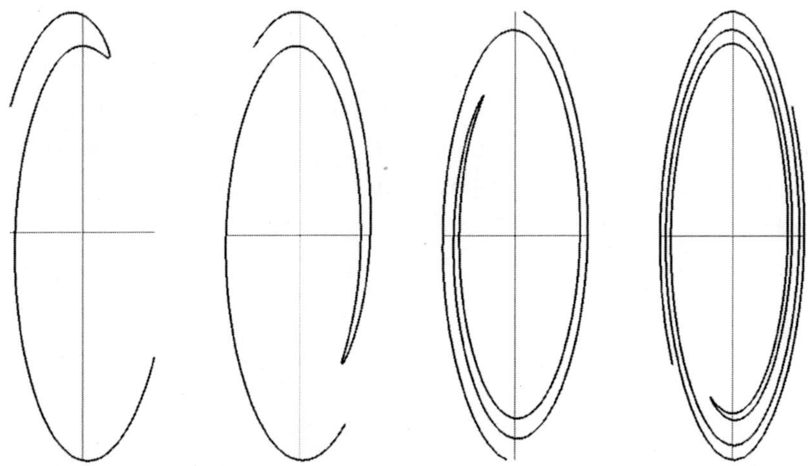

Figure 8. The Present Figure Shows The Evolution Of $\mathcal{L}$ Under The Poincaré Map Φ For Time Intervals Of Different Lengths. The parameters used are k=1$k=1$, μ1=2, μ0=1/2, as before, while for t+−t−, we have considered the cases 35, 60, 100, 150 (from the left to the right). For graphical reasons, a slightly different *x*- and *y*-scaling has been used.

$$w(t_+ - t_-) := \left\lceil \frac{2(t_+ - t_-)}{\tau_{\ell_1}} \right\rceil - \left\lfloor \frac{2(t_+ - t_-)}{\tau_{\ell^*}} \right\rfloor - 2. \tag{2.11}$$

Similarly, the image through Φ of the left part of $\mathcal{L}$ (the sub-arc of $\mathcal{L}$ connecting R to P−) is a spiral-like curve winding a certain number of times in the counterclockwise sense around the origin. Indeed, it connects the points Φ(R) and Φ(P−) inside the region $\mathcal{E}$ (the first point moves at a faster speed than the second one). Again, the number of half-turns of this second curve will depend on w(t+−t−). In the end, gluing together the two spiral-like curves, we conclude that, if t+−t− is sufficiently large, the arc Φ(L) will appear like a double spiral with a central 'hook'. To better describe the structure of Φ(L), we should observe that the 'tip of the hook' is not Φ(R), but it is the image through Φ of a point (close to *R*) on the left part of L\{R} (we owe this remark to the referee). For a formal proof based on the argument outlined above, see Theorem 2.2.

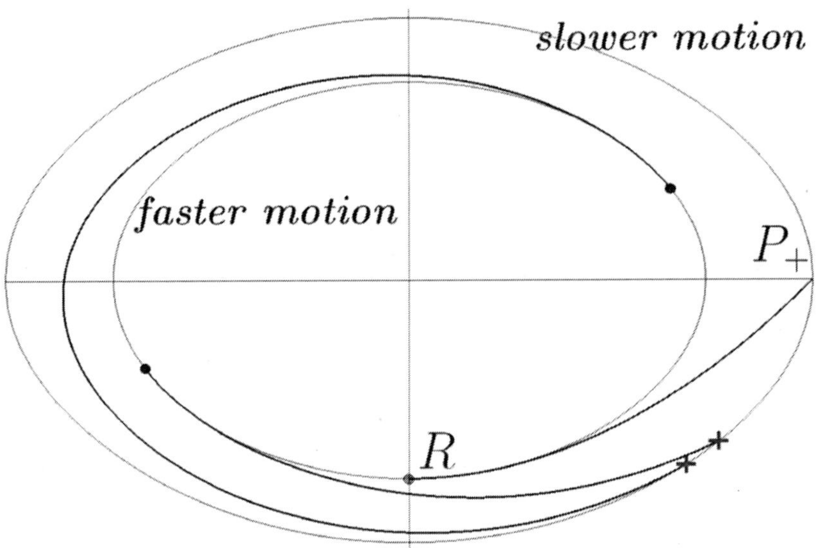

Figure 9. The Present Figure Explains The Twist Effect Due To The Different Periods Of$\gamma_{\ell 1}$and$\gamma_{\ell *}$. We have considered the evolution of the part of $\mathcal{L}$ with x≥0 connecting P+ to the point *R* for two different time gaps: t+−t−=5.5 and t+−t−=50. The corresponding two images Φ(P+) are indicated by a cross, while the images Φ(R) are marked by a black dot. The parameters used are k=1, μ1=2, μ0=1/2, as before. Notice that the two images Φ(P+), although close to each other in the figure, have made a different number of turns around the origin. For graphical reasons, a slightly different *x*- and *y*-scaling has been used.

In preparation for this theorem, we introduce the following notation. Given a positive real number S and an energy level $\ell \in [\ell_1, \ell_*]$, we denote by $\underline{n}(S,\ell) := \lfloor \frac{2S}{\tau_\ell} \rfloor$ the lower integer part of $S/(\tau_\ell/2)$, and by $\overline{n}(S,\ell) := \lceil \frac{2S}{\tau_\ell} \rceil$ the upper integer part of $S/(\tau_\ell/2)$. By definition, during a time interval of length S, a point on the orbit Γ^ℓ makes more that $\overline{n}(S,\ell)-1$ half-turns around the origin, but less that $\underline{n}(S,\ell)+1$ half-turns. Thus the rotational gap defined above can be written as $w(T)=\overline{n}(T,\ell_1)-\underline{n}(T,\ell_*)-2$ for $T=t_+-t_-$. By definition, in the open interval $]2T/\tau_{\ell_*}, 2T/\tau_{\ell_1}[$, the smallest integer is $\underline{n}(T,\ell_*)+1$, while the largest integer is $\overline{n}(T,\ell_*)-1$. Hence, the open interval $]2T/\tau_{\ell_*}, 2T/\tau_{\ell_1}[$ contains $w(T)+1$ positive integers.

2.4 Conclusion

After the preliminary study of the previous subsections, we are now in a position to express our results for the equation

$$x'' + 2kx - 2g(t)x^3 = 0 \qquad 2.1$$

as statements with a formal proof. Recall that the weight function $g(t)$ is defined as follows:

$$g(t) = \begin{cases} \mu_1 & \text{for } t \le t_- \text{ or } t \ge t_+, \\ \mu_0 & \text{for } t_- < t < t_+, \end{cases}$$

Where

$$t_- < t_+ \quad \text{and} \quad 0 < \mu_0 < \mu_1.$$

We also set T:=t+−t− and consider the Poincaré map Φ_0^T associated to the system

$$\begin{cases} x' = y, \\ y' = -2kx + 2\mu_0 x^3. \end{cases} \tag{2.7}$$

In order to make the exposition more clear, we have decided to consider separately the cases of external and internal connections.

THEOREM 2.1 (External connections)

Under the above assumptions, the following results hold for solutions which satisfy the energy condition

$$x'(t)^2 + 2kx(t)^2 - \mu_0 x(t)^4 \in]\ell^*, \ell_0[, \quad \forall t \in \mathbb{R} :$$

- there always exist a *(*positive*)* solution homoclinic to*P+*and a *(*negative*)* solution homoclinic to*P−;*

- if$t_+ - t_- > \tau_{\ell^*}/2$, there always exist a heteroclinic solution fromP_+ toP_- and a heteroclinic solution fromP_- toP_+, both with exactly one zero;

- if$t_+ - t_- > n\tau_{\ell^*}$ (for some integer$n \geq 1$) then, for each integer j with$1 \leq j \leq n$, there exist at least one solution homoclinic to$P_+$and one solution homoclinic toP_-, both with exactly 2j zeros;

- if$t_+ - t_- > (n+1)\tau_{\ell^*}/2$ (for some integer$n \geq 1$) then, for each integer j with$1 \leq j \leq n$, there exist at least one heteroclinic solution fromP_+ to$P_-$and one fromP_- toP_+, both with exactly$2j+1$ zeros.

Proof

As a first step, we focus our attention on the search of solutions homoclinic to the point P_+. The case of homoclinics to P_- is analogous, thus it will be omitted.

Recalling that $T=t_+-t_-$, our goal is to find points on $\mathcal{O}(P_+)$ which are moved by the Poincaré map Φ_0^T on $\mathcal{I}(P_+)$.

We introduce in the phase-plane for (2.7), a system of polar coordinates (θ,ρ), with center in the origin. The initial points in the arc $\mathcal{Y}$ (which is the closure of the intersection of $\mathcal{O}(P_+)$ with the external region $\mathcal{F}$) are parameterized as $\gamma(s):=(\rho(s)\cos\theta(s), \rho(s)\sin\theta(s))$, for $s\in[0,1]$, with $\gamma(0)=P_+$, $\gamma(1)\in\Gamma^{\ell_0}$ and $E(\gamma(s))$ strictly increasing with *s*. The *target set* is the symmetric of $\mathcal{Y}$ with respect to the *x*-axis. It will be denoted by $\mathcal{Y}'$ and parameterized by reversing the angle $\theta(s)$.

Using the same polar coordinates to represent the solutions of (2.7), we can express the final points $\Phi_0^T(\gamma(s))$ as

$$\Phi_0^T(\gamma(s)) = (\rho(T,s)\cos\theta(T,s), \rho(T,s)\sin\theta(T,s)),$$

With

$$\rho(0,s) = \rho(s), \qquad \theta(0,s) = \theta(s).$$

We notice that

$$(x(t), y(t)) = (\rho(t,s)\cos\theta(t,s), \rho(t,s)\sin\theta(t,s))$$

is the solution of (2.7) (at the time t), which departed from the point γ(s)$\gamma(s)$ at the time t=0$t=0$. Observe that, for every s∈[0,1]$s \in [0,1]$, the map

$$t \mapsto \theta(t,s)$$

is strictly decreasing on [0,+∞) (this is an equivalent way to express the fact that the solutions turn around the origin in the clockwise sense).

With these positions, we obtain an external solution homoclinic to P+P_+ if and only if there exists s^∈]0,1[such that $\Phi_0^T(\gamma(\hat{s})) \in \mathcal{Y}'$. This happens if and only if

$$\rho(T,\hat{s}) = \rho(\hat{s}) \quad \text{and} \quad \theta(T,\hat{s}) = -\theta(\hat{s}) - 2j\pi, \tag{2.12}$$

for some nonnegative integer j. In this case, if we denote by x^(⋅) the corresponding homoclinic solution of (2.1) such that (x^(t−),x^′(t−))=γ(s^), we have that E(x^(t),x^′(t))∈]ℓ∗,ℓ0[for all t∈R. Moreover, if j=0, x^(t)>ξ1 for all t∈[t−,t+], while x^(t) has precisely 2j zeros in]t−,t+[if j≥1.

Using the fact that the energy level lines of (2.7) in the region W\{O}={(x,y):−ξ0≤x≤ξ0,0<E(x,y)≤ℓ0} are strictly star-shaped with respect to the origin and symmetric with respect to the x-axis, we find that (2.12) holds (for some nonnegative integer j) if and only if

$$\theta(T,\hat{s}) = -\theta(\hat{s}) - 2j\pi \tag{2.13}$$

is satisfied (for the same j).

By virtue of (2.13), we can refer only to the angular coordinates, hence we will obtain solutions as follows. Since the point $\Phi_0^T(\gamma(1))$ lies on Γℓ0 (which is the trajectory of a heteroclinic solution of (2.7)), it can never reach the point Q+. Accordingly, the angle θ(T,1) satisfies

$$\theta(T,0) < 0, \quad \forall T > 0.$$

On the other hand, as γ(0)=P+, we have that θ(0,0)=θ(0)=0$\theta(0,0) = \theta(0) = 0$, therefore

θ(T,0)<0,∀T>0.

As a consequence, θ(T,0)+θ(0)<0<θ(T,1)+θ(1) and the intermediate value theorem ensures the existence of an s^ such that $\theta(T,\hat{s}) = -\theta(\hat{s})$. In this way we have found a positive homoclinic solution to P+, independently of the length *T* of the time interval [t−,t+].

Suppose now that T>nτℓ∗ for some positive integer *n*. In this case,

$$\theta(T,0) < -2n\pi, \quad \forall T > 0.$$

As a consequence, for every integer *j* with 1≤j≤n, we have θ(T,0)+θ(0)<−2jπ<0<θ(T,1)+θ(1), and again the intermediate value theorem ensures the existence of an s^ such that $\theta(T,\hat{s}) = -\theta(\hat{s}) - 2j\pi$. This ends the proof for homoclinic solutions.

For the search of heteroclinic solutions from P+ to P−, we follow a similar procedure, choosing as a target the set Y″, which is the symmetric of $\mathcal{Y}$ with respect to the *y*-axis (here we exploit the oddness of the nonlinear term x^3). In terms of polar coordinates, the desired solutions will be obtained if and only if there exists s^∈]0,1[$\hat{s} \in]0,1[$ such that

$$\theta(T,\hat{s}) = -\theta(\hat{s}) - (2j-1)\pi \tag{2.14}$$

for some integer j≥1. Then the proof can be concluded as above, via the intermediate value theorem. We observe that the case of heteroclinic solutions from P− to P+ is analogous, thus it will be omitted. □

The next theorem deals with the internal connections. For this result, it is useful to recall the rotational gap defined in (2.11).

THEOREM 2.2 (Internal connections)

Under the above assumptions, the following results hold for solutions which satisfy the energy condition

$$x'(t)^2 + 2kx(t)^2 - \mu_0 x(t)^4 \in [\ell_1, \ell^*], \quad \forall t \in \mathbb{R}:$$

- there always exist a solution homoclinic to*P+*and a solution homoclinic to*P−;*
- there always exist a heteroclinic solution from*P+*to*P−*and a heteroclinic solution from*P−*to*P+;*
- if $t_+ - t_- > \frac{\tau_{\ell_1}\tau_{\ell^*}}{\tau_{\ell^*}-\tau_{\ell_1}}$, there exist at least *2w(t+−t−)+1*solutions homoclinic to *P+*$P_+$and *2w(t+−t−)+1*solutions homoclinic to*P−*P_-*;*
- if $t_+ - t_- > \frac{\tau_{\ell_1}\tau_{\ell^*}}{\tau_{\ell^*}-\tau_{\ell_1}}$, there exist at least *2w(t+−t−)+1*heteroclinic solutions from*P+*$P_+$to*P−*$P_-$and *2w(t+−t−)+1*from *P−* to*P+*P_+.

Proof As the first step, we focus our attention on the search of solutions homoclinic to the point P+. The case of homoclinics to P− is analogous, thus it will be omitted.

Our goal is to find points on $\mathcal{L} = \overline{\mathcal{L}(P_+, P_-)}$ which are moved by the Poincaré map ΦT0Φ_0^T on $\mathcal{L}' = \overline{\mathcal{L}(P_-, P_+)}$.

As a preliminary remark, we note that $\Phi_0^T(\mathcal{L})$ is a simple arc contained in $\mathcal{E}$, connecting its end points $\Phi_0^T(P_+)$ and $\Phi_0^T(P_-)$ on Γ^{ℓ^*} through the 'intermediate' point $\Phi_0^T(R)$ on Γ^{ℓ_1}. Observe also that $\Phi_0^T(P_+)$ and $\Phi_0^T(P_-)$are antipodal. In fact, P− and P+ are antipodal and Φ_0^T is an odd map (indeed, (x(t),y(t)) is a solution of (2.7) if and only if (−x(t),−y(t)) is a solution of the same equation). Now, if the trivial situation $\{\Phi_0^T(P_-), \Phi_0^T(P_+)\} = \{P_-, P_+\}$ occurs, we have $\Phi_0^T(\mathcal{L}) \cap \mathcal{L}' = \{P_-, P_+\} \neq \emptyset$. Otherwise, we find that $\Phi_0^T(P_-)$ and $\Phi_0^T(P_+)$ belong to the two different components of E\L′. Since $\Phi_0^T(P_-)$ and $\Phi_0^T(P_+)$ are the end points of the arc $\Phi_0^T(\mathcal{L}) \subseteq \mathcal{E}$, by an elementary connectivity argument, we conclude that $\Phi_0^T(\mathcal{L}) \cap \mathcal{L}' \neq \emptyset$. This proves the first assertion of the theorem.

For the search of heteroclinic solutions from P+ to P−, we adopt a similar procedure, choosing as a target the set $\mathcal{L}$ itself. The fact that, for every T>0, there is always at least one intersection point between $\Phi_0^T(\mathcal{L})$ and $\mathcal{L}$ follows by the same argument developed for homoclinic solutions. Namely, if the trivial situation $\{\Phi_0^T(P_-), \Phi_0^T(P_+)\} = \{P_-, P_+\}$ occurs, we have $\Phi_0^T(\mathcal{L}) \cap \mathcal{L} = \{P_-, P_+\} \neq \emptyset$. Otherwise, we find that $\Phi_0^T(P_-)$ and $\Phi_0^T(P_+)$ belong to the two different components of E\L. Therefore, it follows that $\Phi_0^T(\mathcal{L}) \cap \mathcal{L}' \neq \emptyset$.

We study now the problem of multiplicity of solutions. As before, we consider at first the case of solutions homoclinic to P+.

We introduce in the inner region $\mathcal{E}$ of a phase-plane for (2.7) a system of modified polar coordinates (θ,E) with center in the origin. In this system, every point is determined by its angular coordinate θ and its energy E defined in (2.8). Since the energy level lines in $\mathcal{E}$ are strictly star-shaped with respect to the origin, we obtain a coordinate system equivalent to the polar one.

Observe that the set L∪L′ intersects every half-line from the origin exactly in one point. Accordingly, we can parameterize those points using the angular coordinate. With this convention, the initial points in the arc $\mathcal{L}$ are parameterized as γ(s) for s∈[π,2π] (*s* is the angle) with γ(π)=P−, γ(2π)=P+ and γ(3π/2)=R. As a consequence, E(γ(s)) is strictly decreasing for s∈[π,3π/2]$s \in [\pi, 3\pi/2]$ and strictly increasing in [3π/2,2π]. The target set L′$\mathcal{L}'$ (the symmetric of $\mathcal{L}$ with respect to the *x*-axis) is parameterized with the angle s∈[0,π]$s \in [0, \pi]$.

For the specific case of system (2.6), an analytic expression for $\mathcal{L}$ as a graph in the (θ,E)-plane, is given as follows:

$$\begin{aligned} E(\theta) &:= E\big(\rho(\theta)\cos\theta, \rho(\theta)\sin\theta\big) \\ &= \rho^2(\theta)\sin^2\theta + 2k\rho^2(\theta)\cos^2\theta - \mu_0\rho^4(\theta)\cos^4\theta, \end{aligned} \tag{2.15}$$

With

$$\rho(\theta) = \frac{(2k/\sqrt{\mu_1})}{(\sin^2\theta + 4k\cos^2\theta)^{1/2} - \sin\theta}, \qquad \theta \in [\pi, 2\pi].$$

Hence, the natural parametrization of $\mathcal{L}$ in the (θ,E)-plane is given by γ(s)=(s,E(s)) for s=θ∈[π,2π].

Using the same modified polar coordinates to represent the solutions of (2.7), we can express the final points $\Phi_0^T(\mathcal{L})$in the (θ,E)-plane by means of their angular coordinate θ(t,s) and energy E(t,s). As in the previous proof, we observe that, for every s∈[π,2π], the map t↦θ(t,s) is strictly decreasing on [0,+∞), while the energy is constant with respect to *t*. Observe also that in the new coordinates (θ,E), system (2.7) becomes

$$\begin{cases} \theta' = -\sin^2\theta - 2k\cos^2\theta + 2\mu_0\Lambda(\theta,E)\cos^4\theta, \\ E' = 0, \end{cases} \tag{2.16}$$

With

$$\Lambda(\theta,E) := \frac{2E}{\sin^2\theta + 2k\cos^2\theta + ((\sin^2\theta + 2k\cos^2\theta)^2 - 4\mu_0 E\cos^4\theta)^{1/2}}.$$

In the (θ,E)-plane, the points of $\mathcal{L}$ move with a negative angular speed along the lines of constant energy. Thus, under the action of the flow associated to (2.16), the points of $\mathcal{L}$ shift from the right to the left in the strip

$$\mathcal{E}(-\infty,2\pi] := \{(\theta,E) : \theta \le 2\pi, \ell_1 \le E \le \ell^*\}.$$

As we have explained before, we obtain an internal solution homoclinic to P+ whenever there is a point z∈L such that $\Phi_0^T(z) \in \mathcal{L}'$, with $T = t_+ - t_-$. In the (θ,E)-plane, this target set is expressed as the union of the graphs

$$\Xi := \bigcup_{j=0}^{\infty}\{(\theta,\Upsilon(\theta)) : -2j\pi \le \theta \le \pi - 2j\pi\},$$

Where

$$\Upsilon(\theta) := E\big(\varrho(\theta)\cos\theta, \varrho(\theta)\sin\theta\big) = \varrho^2(\theta)\sin^2\theta + 2k\varrho^2(\theta)\cos^2\theta - \mu_0\varrho^4(\theta)\cos^4\theta,$$

With

$$\varrho(\theta) = \frac{(2k/\sqrt{\mu_1})}{(\sin^2\theta + 4k\cos^2\theta)^{1/2} + \sin\theta}$$

(see Figures 10, 11 for a graphical description).

In this setting, we obtain an internal solution homoclinic to P+P_+ if and only if there exists s^∈[π,2π] such that Φ(γ(s^))∈Ξ, where we have denoted by Φ the Poincaré map for (2.16) in the time interval [0,T]. Of course Φ is exactly Φ_0^T (which was the Poincaré map associated to (2.7)) in the new (θ,E)-coordinates. Using the parameterized curve γ(s) for the initial points, we can express Φ as

$$\Phi\big(\gamma(s)\big) = \big(\theta(T,s), E(s)\big) \quad \text{for } s \in [\pi, 2\pi]$$

(recall that γ(s)=(s,E(s)) for s=θ∈[π,2π]).

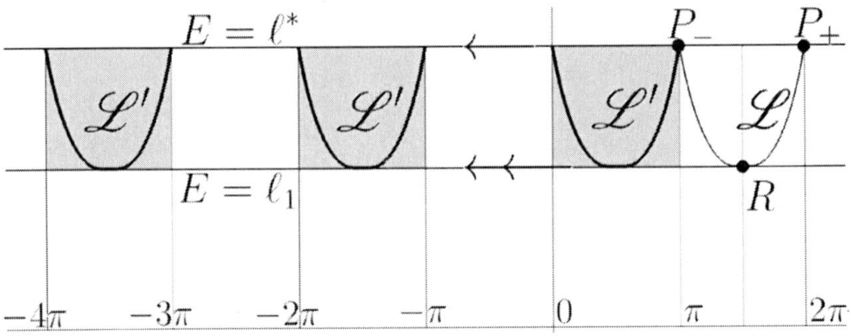

Figure 10. The Initial Set $\mathcal{L}$ And The Target Set$\mathcal{l}'\mathcal{L}'$Represented In The(θ,E)(θ,E)-Plane. The present figure is drawn for the parameters k=1$k=1$, μ1=2$\mu_1=2$, μ0=1/2$\mu_0=1/2$. The internal region $\mathcal{E}$ corresponds to the strip R×[ℓ1,ℓ∗]. Since we are interested in the evolution of the set $\mathcal{L}$through the Poincaré map, we consider only the half-strip E(−∞,2π]=(−∞,2π]×[ℓ1,ℓ∗]. For the flow associated to (2.16), all the points of $\mathcal{L}$ move from the right to the left on lines parallel to the θ-axis. The point R (as well as the points on the line E=ℓ1$E=\ell_1$) moves faster than the points P± which are on the line E=ℓ∗. During all the evolution, the distance between the images of P− and P+ remains constantly equal to π.

The points P− and P+ are antipodal, lie on the same energy line Γℓ∗ and move with the same angular speed. In the (θ,E)-plane, P− and P+ are expressed by P−=(π,ℓ∗), P+=(2π,ℓ∗). Therefore,

$$\theta(t,\pi) - \theta(t,2\pi) = -\pi, \quad \forall t \geq 0. \tag{2.17}$$

Moreover, for every integer j≥0, we have that

$$\theta(t,\pi) \lesseqgtr \pi - j\pi \quad \text{if and only if} \quad t \gtreqless j\frac{\tau_{\ell^*}}{2}$$

(this follows from the fact that the orbit Γℓ∗ has a period τℓ∗ and is symmetric with respect to the *x*-axis).

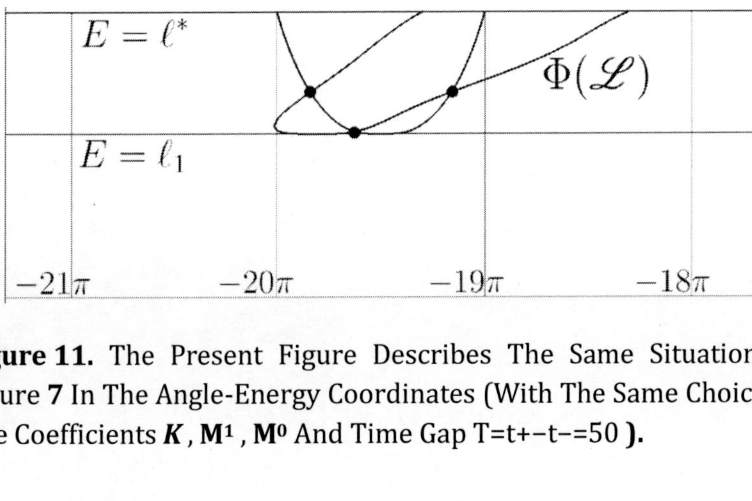

Figure 11. The Present Figure Describes The Same Situation Of Figure **7** In The Angle-Energy Coordinates (With The Same Choice Of The Coefficients ***K***, **M¹**, **M⁰** And Time Gap T=t+−t−=50 **).**

In the same plane, the point *R* is indicated by R=(3π/2,ℓ1). Using the fact that the orbit $\Gamma\ell 1\Gamma^{\ell_1}$ has a period τℓ1 and is symmetric with respect to the *y*-axis, we find that, for every integer i≥0,

$$\theta(t,3\pi/2) \lesseqgtr \frac{\pi}{2} - 2i\pi \quad \text{if and only if} \quad t \gtreqless \frac{\tau_{\ell_1}}{2} + i\tau_{\ell_1}.$$

Suppose now that

$$\frac{2T}{\tau_{\ell_1}} - \frac{2T}{\tau_{\ell^*}} > 2. \tag{2.18}$$

Then, the open interval $]2T/\tau_{\ell*},2T/\tau_{\ell_1}[$ contains at least one odd integer.

Let $m_o(T)$ be the largest odd integer contained in $]2T/\tau_{\ell*},2T/\tau_{\ell_1}[$. From $T > \mathbf{m}_o(T)\tau_{\ell_1}/2$, we have that

$$\theta(T,3\pi/2) < \frac{\pi}{2} - (\mathbf{m}_o(T)-1)\pi = \frac{3\pi}{2} - \mathbf{m}_o(T)\pi. \tag{2.19}$$

Suppose also that j is a positive integer with $2T/\tau\ell* < j \le mo(T)$. In this case, from $T < j\tau\ell*/2$, we have that

$$\theta(T,\pi) > \pi - j\pi \quad \text{and} \quad \theta(T,2\pi) > 2\pi - j\pi. \tag{2.20}$$

Let

$\mathcal{L}_{\text{left}} := \{(\theta,E(\theta)) : \pi \le \theta \le 3\pi/2\}$ and

$\mathcal{L}_{\text{right}} := \{(\theta,E(\theta)) : 3\pi/2 \le \theta \le 2\pi\}$,

with $\mathcal{L} = \mathcal{L}_{\text{left}} \cup \mathcal{L}_{\text{right}}$ be the left and the right parts of $\mathcal{L}$ in the (θ,E)-plane. Similarly, we define

$\mathcal{L}'_{\text{left}} := \{(\theta,\Upsilon(\theta)) : 0 \le \theta \le \pi/2\}$

And

$$\mathcal{L}'_{\text{right}} := \{(\theta, \Upsilon(\theta)) : \pi/2 \leq \theta \leq \pi\}.$$

By definition, for any nonnegative integer *i*, the set L′ $\mathcal{L}'_{\text{left}} - (2i\pi, 0)$ is a simple arc connecting the points (−2iπ,ℓ∗) and (−2iπ+(π/2),ℓ1) in the rectangle [−2iπ,−2iπ+(π/2)]×[ℓ1,ℓ∗] and, similarly, $\mathcal{L}'_{\text{right}} - (2i\pi, 0)$ is a simple arc connecting the points (−2iπ+(π/2),ℓ1) and (−2iπ+π,ℓ∗) in the rectangle [−2iπ+(π/2),−2iπ+π]×[ℓ1,ℓ∗]. On the other hand, the set Φ(Lleft) is a simple arc connecting the points (θ(T,π),ℓ∗) and (θ(T,3π/2),ℓ1) in E(−∞,2π]$\mathcal{E}(-\infty, 2\pi]$ and, similarly, Φ(Lright) is a simple arc connecting the points (θ(T,3π/2),ℓ1) and (θ(T,2π),ℓ∗)=(π+θ(T,π),ℓ∗) in the same strip.

Suppose now that j=mo(T)=2k+1 (in this case, *j* is odd). In such a situation, from (2.19) and (2.20) we have that

$$\theta(T, 3\pi/2) < \frac{\pi}{2} - 2k\pi, \qquad \theta(T, \pi) > -2k\pi, \qquad \theta(T, \pi) > \pi - 2k\pi,$$

for 2k=j−1 an even integer. Hence the points (θ(T,3π/2),ℓ1) and (θ(T,π),ℓ∗) are separated by the arc L′ $\mathcal{L}'_{\text{left}} - (2k\pi, 0)$, while the points (θ(T,3π/2),ℓ1) and (θ(T,2π),ℓ∗) are separated by $\mathcal{L}'_{\text{left}} - (2k\pi, 0)$, and L′right−(2kπ,0). As a consequence, Φ(L)∩Ξ contains at least three points, precisely the nonempty intersections of , Φ(Lright)∩(L′left−(2kπ,0))

Suppose that j=mo(T)−1 (in this case, *j* is even). In such a situation, from (2.19) and (2.20) we have that

$$\theta(T,3\pi/2) < \frac{\pi}{2} - 2k\pi, \qquad \theta(T,\pi) > \pi - 2k\pi, \qquad \theta(T,\pi) > 2\pi - 2k\pi$$

for 2k=mo(T)−1 an even integer. Hence the points (θ(T,3π/2),ℓ1) and (θ(T,π),ℓ∗) are separated by the arcs $\mathcal{L}'_{\text{left}} - (2k\pi,0)$, $\mathcal{L}'_{\text{right}} - (2k\pi,0)$ and $\mathcal{L}'_{\text{left}} - ((2k-1)\pi,0)$. The points (θ(T,3π/2),ℓ1) and (θ(T,2π),ℓ∗) are separated by $\mathcal{L}'_{\text{left}} - (2k\pi,0)$, $\mathcal{L}'_{\text{right}} - (2k\pi,0)$ and $\mathcal{L}'_{\text{left}} - ((2k-1)\pi,0)$. $\mathcal{L}'_{\text{left}} - ((2k-1)\pi,0)$. As a consequence, Φ(L)∩Ξ contains at least five points: two points coming from the intersections of Φ(L_{left})with the target set Ξ and three from the intersections of Φ(L_{right}).

Proceeding by induction, with the same argument, the following result is obtained.

CLAIM 1 Φ(L)∩Ξ *has at least* 2n#+1 *solutions, where* n# *is the number of integers less than or equal to* mo(T) *which are contained in the open interval*]2T/τℓ∗,2T/τℓ1[.

A completely symmetric argument leads to the same multiplicity result for solutions which are homoclinic to P−.

At last, we look for a multiplicity result for heteroclinic solutions from P+ to P−. We follow step by step the argument described in the part of the proof devoted to the search of multiple internal homoclinic solutions and, therefore, we transform our equation in the (θ,E)-coordinates. As we have explained before, we obtain an internal heteroclinic (from P+ to P−) whenever there is a point z∈L such that $\Phi_0^T(z) \in \mathcal{L}$, with T=t+−t−. In the (θ,E)-plane, this target set is expressed as the union of the graphs

$$\Xi' := \bigcup_{j=0}^{\infty}\{(\theta, E(\theta)) : \pi - 2j\pi \leq \theta \leq 2\pi - 2j\pi\},$$

where E(θ) is the function defined in (2.15).

As before, we assume the validity of condition (2.18). Then the open interval $]2T/\tau_{\ell*}, 2T/\tau_{\ell 1}[$ contains at least one even integer.

Let me(T) be the largest even integer contained in $]2T/\tau_{\ell*}, 2T/\tau_{\ell 1}[$. From $T > m_e(T)\tau_{\ell 1}/2$, we have that

$$\theta(T, 3\pi/2) < \frac{3\pi}{2} - \mathbf{m}_e(T)\pi.$$

For any positive integer *j* with $2T/\tau_{\ell*} < j \leq m_e(T)$, we have that (2.20) holds too. At this point, we have simply to repeat (with obvious changes due to the fact that the target set is Ξ′Ξ′ which is a shift of Ξ by (2pi,0)$(2pi,0)$) the argument presented above for the case of homoclinic connections and obtain the following conclusion.

CLAIM 2 Φ(L)∩Ξ′ *has at least* 2n#+1 *solutions, where* n# *is the number of integers less than or equal to* $m_e(T)$ *which are contained in the open interval* $]2T/\tau_{\ell*}, 2T/\tau_{\ell 1}[$.

A completely symmetric argument leads to a multiplicity result for heteroclinic solutions from P− to P+.

As final step, we have just to show how the condition on the rotational gap allows us to achieve the conclusion from Claim 1 and Claim 2 that we have obtained along the proof. Now, it is sufficient to observe that $t_+ - t_- = T > \frac{\tau_{\ell_1}\tau_{\ell^*}}{\tau_{\ell^*}-\tau_{\ell_1}}$ implies (2.18) and, therefore, the open interval $]2T/\tau_{\ell^*}, 2T/\tau_{\ell_1}[$. contains at least two integers. More precisely, by the definition of $w(T)$, the interval $]2T/\tau_{\ell^*}, 2T/\tau_{\ell_1}[$. contains at least w(T)+1$w(T)+1$ integers and, therefore,

$$n^{\#}, n_{\#} \geq w(T).$$

This concludes the proof of the theorem.

3 REMARKS AND RELATED RESULTS

We end the paper with a list of remarks about possible variants and extensions of the main results obtained for equation

$$x'' + 2kx - 2g(t)x^3 = 0. \tag{3.1}$$

- A problem which naturally arises from the analysis that we have performed concerns what happens if, for a stepwise weight function satisfying (2.5), we suppose that $0<\mu_1<\mu_0$. Repeating the preliminary phase-plane analysis of Section 2.2, one can easily check that for any gap T=t+−t−>0, there always exist a solution homoclinic to P+ and another homoclinic to P−P_-, as well as a heteroclinic from P+ to P− and another one from P−P_- to P+P_+. However, in general, if 0<μ1<μ0, one cannot obtain multiplicity results like those achieved in Section 2 without some extra assumptions on h(x). This is the reason for which, in the study of equation (2.1), we have considered only the case 0<μ0<μ1. For different examples on related equations in which a weight coefficient can be above or below its limits at infinity, see, for instance, [4–6].

- On the other hand, if we assume that the weight function changes its sign, with μ0<0<μ1, some interesting multiplicity results could be produced. In fact the phase portrait of (1.1) in the time interval [t−,t+] shows a global center, hence if t+−t− is large, a lot of connections between the unstable and stable manifolds of the points P± can be obtained. For the sake of conciseness, we omit the study of this latter situation, which is beyond the goal of the present paper.

- We notice that the same argument of the proofs applies to an equation of the form

$$x'' + \sigma x - a(t)h(x) = 0 \tag{3.2}$$

for σ>0, h(x)=xf(x), and with f(x)$f(x)$ satisfying condition (∗), if we assume that a(t) is a stepwise weight function, playing the same role of g(t)$g(t)$ in (2.1). In such a case, the saddle points become P±=(±ξ1,0) with ξ1:=f-1(σ/μ1), where f−1f^{-1} is the inverse of *f*restricted to [0,+∞). In this way, we can apply our results to nonlinearities like those considered in [2, 27]. More precisely, Theorem 2.1 holds without any further assumption, while for Theorem 2.2 we need to require a gap between the periods of the orbits Γℓ1Γ^{ℓ_1} and Γℓ∗Γ^{ℓ^*}. In order to obtain this gap, we can apply (for instance) some results ensuring the monotonicity of the time-map (like [[28], Theorem A]).

Due to the special form of the weight coefficient, it is standard to verify (via a simple rescaling procedure) that (1.1) is equivalent to an equation of the form

$$x'' + b(t)x - ah(x) = 0, \tag{3.3}$$

where a>0 and b(t) is a stepwise coefficient. In this manner, one can deal with some nonlinear Schrödinger-type equations related to the case of potential wells or potential walls [29, 30].

The approach used in the proofs, based on the properties of the Poincaré map, guarantees that our results are stable with respect to small perturbations. More precisely, fixed a suitable length T=t+−t−$T = t_+ - t_-$ for the time interval, Theorems 2.1 and 2.2 provide a lower bound for the number of solutions. We can state that the same lower bound persists for a small perturbation of the coefficient in the L1L^1-norm on [t−,t+]. Therefore, the assumption that the weight a(t)$a(t)$ in (1.1) or b(t) in (3.1) are stepwise functions can be slightly relaxed, so we can 'smooth' them.

With reference to equations (1.1) or (3.1) with stepwise coefficients, we observe that our approach can be adapted to boundary value problems on a compact interval [0,T] like, *e.g.*, the Dirichlet (two-point) or the Neumann problem. In these cases, we have to find solutions connecting given lines which depart from the origin. For the sake of conciseness, we cannot describe the most general situation, but we just outline a possible application for the Neumann problem

$$\begin{cases} x'' + 2kx - 2g(t)x^3 = 0, \\ x'(0) = 0, \qquad x'(T) = 0, \end{cases} \tag{3.4}$$

with $g(t)$$g(t)$ a stepwise function such that $g(t)=\mu_1$ for $t \in [0,\tau[$ and $g(t)=\mu_0$ for $t \in [\tau,T]$, for $0<\mu_0<\mu_1$. Let us denote by Φ_1 the Poincaré map associated to system (2.6) on the time interval $[0,\tau]$$[0,\tau]$, and by Φ_2 the Poincaré map associated to system (2.7) on the time interval $[0,T-\tau]$. We observe that there exists a maximal compact interval $[-\xi^*_1,\xi^*_1]$$[-\xi_1^*,\xi_1^*]$ with $\xi_1<\xi^*_1<\xi_0$ such that $E(\Phi_1(z))\leq \ell_0$$E(\Phi_1(z)) \leq \ell_0$ for all the points $z=(x,0)$$z=(x,0)$ with $x\in[-\xi^*_1,\xi^*_1]$. In this manner $\Phi_1(\pm\xi^*_1,0)\in\Gamma\ell_0$. The curve

$$\Upsilon^* := \left\{ \Phi_1(x,0) : -\xi_1^* \leq x \leq \xi_1^* \right\}$$

represents the set of all the points in the region $W=\{(x,y):-\xi_0\leq x\leq \xi_0, E(x,y)\leq \ell_0\}$, which are images (by Φ) of initial points of the *x*-axis (therefore $x'(0)=0$). Figure 12 illustrates the curve $\Upsilon*$ for a short time interval ($\tau=2.2$$\tau=2.2$). For a larger τ, the line $\Upsilon*$ becomes a double spiral with a certain number of turns around the origin, while the part of $\Upsilon*$ contained in the region $\mathcal{F}$ gets very close to $O(P+)$ and $O(P-)$. Then we can repeat the same argument developed in the previous sections by looking for the intersections of $\Phi_2(\Upsilon*)$ with the *x*-axis.

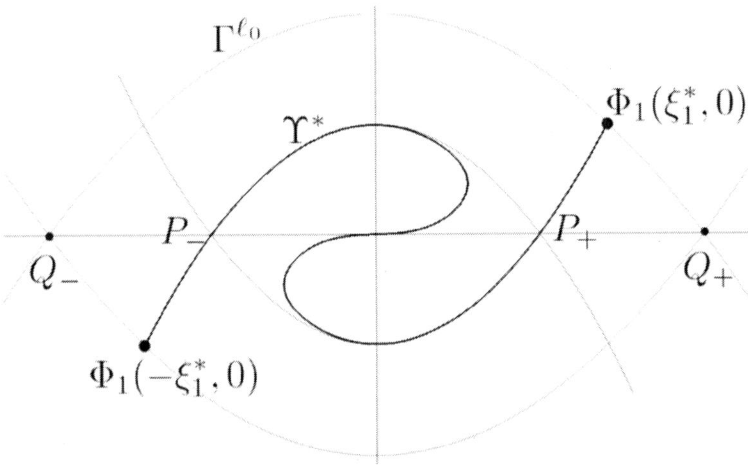

Figure 12. In The Present Figure The Line**υ∗ϒ***Is The Transformation Of The Segment**[−Ξ∗1,Ξ∗1]×{0}**Under The Poincaré Map**φ1**for**τ**=2.2. The parameters used are k=1$k=1$, μ1=2$\mu_1=2$, μ0=1/2$\mu_0=1/2$. For graphical reasons, a slightly different *x*- and *y*-scaling has been used.

As a final remark, we mention the fact that combining our technique with Ważewski's method [31], following the approach developed by Conley in [18], one can deal with some more general classes of weight functions. For example, one could tackle with these techniques the cases of *asymptotically constant* or *asymptotically periodic* coefficients. These extensions, however, need a more delicate analysis and they are beyond the goals of the present paper.

We hope that the abundance of multiplicity results found in the present work (in the special case of stepwise coefficients) may suggest possible directions for extending Theorem 2.1 and Theorem 2.2 to more general weight functions. This will be our goal for a future investigation of the problem.

ACKNOWLEDGEMENTS

The authors are deeply indebted with the referee for the careful checking of the manuscript and for his/her remarks, including a correction to an erroneous argument in the previous version of the proof. This research was partially supported by the project PRIN-2009 *Equazioni Differenziali Ordinarie e Applicazioni*.

COMPETING INTERESTS

The authors declare that they have no competing interests.

AUTHORS' CONTRIBUTIONS

All authors contributed equally to the manuscript and read and approved the final manuscript.

REFERENCES

1. Belmonte-Beitia J, Torres PJ: Existence of dark soliton solutions of the cubic nonlinear Schrödinger equation with periodic inhomogeneous nonlinearity. J. Nonlinear Math. Phys. 2008, 15(3):65–72.

2. Torres PJ, Konotop VV: On the existence of dark solitons in a cubic-quintic nonlinear Schrödinger equation with a periodic potential.Commun. Math. Phys. 2008, 282: 1–9.

3. Ambrosetti A, Badiale M, Cingolani S: Semiclassical states of nonlinear Schrödinger equations. Arch. Ration. Mech. Anal. 1997, 140: 285–300.

4. Gavioli A: On the existence of heteroclinic trajectories for asymptotically autonomous equations. Topol. Methods Nonlinear Anal. 2009, 34: 251–266.

5. Gavioli A: Monotone heteroclinic solutions to non-autonomous equations via phase plane analysis. Nonlinear Differ. Equ. Appl. 2011, 18: 79–100.

6. Gavioli A, Sanchez L: Heteroclinics for non-autonomous second-order differential equations. Differ. Integral Equ. 2009, 22: 999–1018.

7. Holmes PJ, Stuart CA: Homoclinic orbits for eventually autonomous planar flows. Z. Angew. Math. Phys. 1992, 43: 598–625.

8. Jones CKRT, Küpper T, Schaffner K: Bifurcation of asymmetric solutions in nonlinear optical media. Z. Angew. Math. Phys. 2001, 52: 859–880.

9. Aronson DG, Mantzaris NV, Othmer HG: Wave propagation and blocking in inhomogeneous media. Discrete Contin. Dyn. Syst. 2005, 13: 843–876.

10. Lewis TJ, Keener JP: Wave-block in excitable media due to regions of depressed excitability. SIAM J. Appl. Math. 2000, 61: 293–316.

11. Yang J, Kalliadasis S, Merkin JH, Scott SK: Wave propagation in spatially distributed excitable media. SIAM J. Appl. Math. 2002, 63: 485–509

12. Rodrigues AS, Kevrekidis PG, Porter MA, Frantzeskakis DJ, Schmelcher P, Bishop AR: Matter-wave solitons with a periodic, piecewise-constant scattering length. Phys. Rev. A 2008., 78: Article ID 013611

13. Margheri A, Rebelo C, Zanolin F: Connected branches of initial points for asymptotic BVPs, with application to heteroclinic and homoclinic solutions. Adv. Nonlinear Stud. 2009, 9: 95–135.

14. Boscaggin A, Zanolin F: Pairs of nodal solutions for a class of nonlinear problems with one-sided growth conditions. Adv. Nonlinear Stud.2013, 13: 13–55.

15. Capietto A, Henrard M, Mawhin M, Zanolin F: A continuation approach to some forced superlinear Sturm-Liouville boundary value problems. Topol. Methods Nonlinear Anal. 1994, 3: 81–100.

16. Fonda, A, Garrione, M: Generalized Sturm-Liouville boundary conditions for first order differential systems in the plane. Topol. Methods Nonlinear Anal. (to appear)

17. Rebelo C, Zanolin F: On the existence and multiplicity of branches of nodal solutions for a class of parameter-dependent Sturm-Liouville problems via the shooting map. Differ. Integral Equ. 2000, 13: 1473–1502.

18. Conley C: An application of Wazewski method to a non-linear boundary value problem which arises in population genetics. J. Math. Biol.1975, 2: 241–249.

19. Rebelo C, Sanchez L: Existence and multiplicity for an O.D.E. with nonlinear conditions. Differ. Equ. Dyn. Syst. 1995, 3: 383–396.

20. Struwe M: Multiple solutions of anticoercive boundary value problems for a class of ordinary differential equations of second order. J. Differ. Equ. 1980, 37: 285–295.

21. Gaines RE, Mawhin J: Ordinary differential equations with nonlinear boundary conditions. J. Differ. Equ. 1977, 26: 200–222.

22. Gaines RE, Mawhin J Lecture Notes in Mathematics 568. In Coincidence Degree, and Nonlinear Differential Equations. Springer, Berlin; 1977.

23. Hale JK: Ordinary Differential Equations. 2nd edition. Krieger, Huntington; 1980.

24. Zanini C, Zanolin F: An example of chaos for a cubic nonlinear Schrödinger equation with periodic inhomogeneous nonlinearity. Adv. Nonlinear Stud. 2012, 12: 481–499.

25. Gradshteyn IS, Ryzhik IM: Table of Integrals, Series, and Products. 7th edition. Academic Press, Boston; 2007. Translation Edited and with a Preface by Alan Jeffrey

26. Meyer KR: Jacobi elliptic functions from a dynamical systems point of view. Am. Math. Mon. 2001, 108: 729–737.

27. Belmonte-Beitia J, Cuevas J: Existence of dark solitons in a class of stationary nonlinear Schrödinger equations with periodically modulated nonlinearity and periodic asymptotics. J. Math. Phys. 2011., 52: Article ID 032702

28. Chicone C: The monotonicity of the period function for planar Hamiltonian vector fields. J. Differ. Equ. 1987, 69: 310–321.

29. Frauenkron H, Grassberger P: Interaction of nonlinear Schrödinger solitons with an external potential. J. Phys. A 1995, 28: 4987–4996.

30. Sakaguchi H, Tamura M: Scattering and trapping of nonlinear Schrödinger solitons in external potentials. J. Phys. Soc. Jpn. 2004, 73: 503–506.

31. Ważewski T: Sur un principe topologique de l'examen de l'allure asymptotique des intégrales des équations différentielles ordinaires. Ann. Soc. Polon. Math. 1947, 20: 279–313.

CHAPTER 8

About reducing integro-differential equations with infinite limits of integration to systems of ordinary differential equations

Yakov Goltser and Alexander Domoshnitsky*

ABSTRACT

The purpose of this paper is to propose a method for studying integro-differential equations with infinite limits of integration. The main idea of this method is to reduce integro-differential equations to auxiliary systems of ordinary differential equations.

Results

A scheme of the reduction of integro-differential equations with infinite limits of integration to these auxiliary systems is described and a formula for representation of bounded solutions, based on fundamental matrices of these systems, is obtained.

Conclusion

Methods proposed in this paper could be a basis for the Floquet theory and studies of stability, bifurcations, parametric resonance and various boundary value problems. As examples, models of tumor-immune system interaction, hematopoiesis and plankton-nutrient interaction are considered.

KEYWORDS

integro-differential equations; fundamental matrix; Cauchy matrix; hyperbolic systems

1 INTRODUCTION

Integro-differential equations appeared very naturally in various applications (see, for example, [1–5]), which explains the interest in the theory of these equations (see, for example, [6, 7]). Various examples, in which the simple enough integro-differential equation

$$x'(t) = X\left(t, x(t), \int_0^t F(t, s, x(s))\, ds\right), \tag{1.1}$$

by elementary operations can be reduced to a system of ordinary differential equations, are known. In this connection, let us refer, for example, to the monograph [8]. Note the idea of the chain trick used in various applications (see, for example, [9, 10]) and its developed form in the paper [11]. Independently, the idea of a reduction to systems of ordinary differential equations in the study of stability, which was, actually, the chain trick, was presented in [12]. Starting with this reduction, approaches to the study of stability and bifurcation of integro-differential equations were proposed in the papers [13–16]. The approach developed in these papers allowed researchers to define a notion of periodic integro-differential systems and to build the Floquet theory for integro-differential equations on this basis in [17]. The first known results on estimates of distance between two adjacent zeros of oscillating solutions to a linearization of equation (1.1) and results connecting oscillation behavior and the exponential stability were obtained on this basis [17]. A parametric resonance in linear almost periodic systems was studied in [18], and the bifurcation of steady resonance modes for integro-differential systems was investigated in [19]. Stabilization by control in a form of integrals of solutions was studied in [20]. The stability of partial functional differential equations on the basis of this reduction was studied in [21]. Constructive approach to a phase transition model was presented in [22]. A reduction to infinite dimensional systems was considered in [21, 23, 24]. In all these papers the limits of integration in integral terms were 0 and *t*, and this was very essential.

The main goal of this paper is to present a method reducing integro differential equations with infinite limits of integration

$$x'(t) = f\left(t, x(t), \int_{-\infty}^{+\infty} K(t,s)g(s,x(s))\,ds\right), \quad t \in (-\infty, +\infty), \tag{1.2}$$

to systems of ordinary differential equations. In a future we are planning to develop the ideas of noted above papers for equation (1.2). As well as we know, there are no results of this type. Important motivation in the study of integro-differential equation (1.2) can be found also in various applications of such equations in, for example, models of tumor-immune system interaction [9], hematopoiesis [10], stability and persistence in plankton models [25] which will be considered below.

Denote

$$u(t) = \int_{-\infty}^{+\infty} K(t,s)g(s,x(s))\,ds, \tag{1.3}$$

$$v(t) = \int_{-\infty}^{t} K(t,s)g(s,x(s))\,ds, \tag{1.4}$$

$$w(t) = \int_{t}^{+\infty} K(t,s)g(s,x(s))\,ds. \tag{1.5}$$

Using these notations, we can write

$$x'(t) = f(t, x(t), u(t)), \quad t \in (-\infty, +\infty), \tag{1.6}$$

Or

$$x'(t) = f\left(t, x(t), v(t) + w(t)\right), \quad t \in (-\infty, +\infty).$$

It is possible to represent the vector $x \in R^n$ in the form $x = col\{y, z\}$, where $y \in Rk$, $z \in R^{n-k}$. In many applications, system (1.4) can be represented in the form

$$y'(t) = Y\left(t, x(t), v(t)\right), \qquad z'(t) = Z\left(t, x(t), w(t)\right), \quad t \in (-\infty, +\infty).$$

(1.7)

The first equation in (1.7) depends on its integral part v on delay only (see (1.4)) and the second one is dependent on advance only. Note that the cases k=n and k=0 can be also considered. If k=n, we get a system with distributed delay, and if k=0, the one with distributed advance. Note that a combination of distributed and concentrated deviations is also possible. Considering such systems, we do not discuss questions of existence of solutions and assume that solutions to these systems exist. Note that even for the Volterra equation, one-point problem (1.1) with the condition x(t0)=x0, t0>0$t_0 > 0$, can have more than one solution or not have solutions at all (see, for example, [26], Chapter 1, Section 9, pp. 70-74).

For system (1.2) our method essentially uses the properties of linear nonhomogeneous systems of ODEs, possessing exponential dichotomy [27] or hyperbolicity [28]. It is known that such systems have (under corresponding conditions) unique bounded on the axis solution. Corresponding bibliography can be found in [28]. The case of autonomous systems was considered in [29, 30]. Below, in the next paragraph, we formulate, in convenient for us form, a result about the existence and structure of the solution for general non-autonomous linear systems of ODEs. This result is based on the theorem about reduction of hyperbolic systems to a block diagonal form [28].

2 METHODS: ABOUT BOUNDED SOLUTIONS OF LINEAR NONHOMOGENEOUS SYSTEMS

Consider

$$x'(t) = P(t)x(t) + g(t), \quad t \in (-\infty, +\infty) \tag{2.1}$$

and the corresponding homogeneous system

$$w'(t) = P(t)w(t), \quad t \in (-\infty, +\infty), \tag{2.2}$$

where $x, w \in R^n$, P is an $n \times n$ matrix and g is an n-vector function with continuous bounded elements.

We use the following definition introduced in [28].

DEFINITION 2.1 We say that system (2.2) is hyperbolic if there exist constants a>0$a > 0$ and λ>0$\lambda > 0$ and hyperplanes M+ and M−: dim M+=k, dimM−=n−k such that if for t=t0$t = t_0$, w(t0)=w0∈M+, then the solution w(t,t0,w0) satisfies the inequality

$$\left|w(t,t_0,w_0)\right| \le a\left|w_0\right|e^{-\lambda(t-t_0)}, \quad t \ge t_0, \tag{2.3}$$

and if w0∈M−, the inequality

$$\left|w(t,t_0,w_0)\right| \le a\left|w_0\right|e^{\lambda(t-t_0)}, \quad t \le t_0. \tag{2.4}$$

THEOREM 2.1 [28]

Let system *(2.2)* be hyperbolic. Then there exists an *n×n* matrix *U(t)*$U(t)$ with bounded elements such that its inverse matrix *U−1(t)*$U^{-1}(t)$ also possesses bounded elements and the transform *w=U(t)η* reduces system *(2.2)* to the form

$$\xi'(t) = Q^{+}(t)\xi(t), \qquad \zeta'(t) = Q^{-}(t)\zeta(t), \tag{2.5}$$

where *η=col{ξ,ζ}*, *ξ∈*R^k, *ζ∈*R^{n-k}.

If we denote *Φ+(t,s)=ϕ+(t)*ϕ^{-1}*+(s)*, where *ϕ+(t)* is a fundamental matrix of the first system in *(2.5)*, *Φ−(t,s)=ϕ−(t)*ϕ^{-1}*−(s)*, where *ϕ−(t)* is a fundamental matrix of the second system in *(2.5)*, such that *Φ+(s,s)=E+*, *Φ−(s,s)=−E−*, *dimE+=k*, *dimE−=n−k*, then

$$\left\|\Phi_+(t,s)\right\| \le ae^{-\lambda(t-s)}, \quad t \ge s, \tag{2.6}$$

$$\left\|\Phi_-(t,s)\right\| \le ae^{\lambda(t-s)}, \quad t \le s. \tag{2.7}$$

We present corresponding constructions, developed in [28] for the proof of this theorem, which will be used below in our paper.

Let

$$\Phi(t) = \left(w_1(t), \ldots, w_n(t)\right) \tag{2.8}$$

be a fundamental matrix of system (2.2), where wi(t) (i=1,...,n) are linearly independent solutions of system (2.2), M+= span(w1,...,wk(t)), M−=span(wk+1,...,wn(t)).

Setting $v_1(t) = w_1(t)$, $u_1(t) = \frac{w_1(t)}{\|w_1(t)\|}$, we define, for m=2,3,...,k, the vectors

$$v_m = w_m - \sum_{i=1}^{m-1}(w_m, u_i)u_i, \qquad u_m = \frac{v_m}{\|v_m\|}. \tag{2.9}$$

For m=k+1, we set $v_{k+1}(t) = w_{k+1}(t),\ u_{k+1}(t) = \frac{w_{k+1}(t)}{\|w_{k+1}(t)\|}$, and for m=k+2,...,n, we define corresponding vectors according to scheme (2.9). The matrix

$$U(t) = \left(u_1(t), \ldots, u_n(t)\right) \tag{2.10}$$

is bounded with its inverse matrix $U^{-1}(t)$ and $\frac{dU}{dt}$. The vectors uj(t) are pairwise orthogonal and ‖uj(t)‖=1, j=1,...,n$j = 1, \ldots, n$. Let us set

$$U(t) = \Phi(t)S(t). \tag{2.11}$$

It is clear from the construction of the matrix U(t) that S(t) is a block diagonal

$$S(t) = \operatorname{diag}\left(S_+(t), S_-(t)\right), \qquad \dim S_+ = k, \qquad \dim S_- = n - k. \tag{2.12}$$

Setting in (2.2) w=U(t)η, we get

$$\frac{d\eta}{dt} = Q(t)\eta, \tag{2.13}$$

where $Q(t) = U^{-1}(PU - \frac{du}{dt}] = -S^{-1}\frac{dS}{dt}$ It follows from (2.12) and (2.13) that Q(t)=diag(Q+(t),Q−(t)), where $Q_+(t) = S_+\frac{dS_+}{dt}, Q_-(t) = S_-\frac{dS_-}{dt}$ Thus system (2.13) has the form

$$\frac{d\xi}{dt} = Q_+(t)\xi, \qquad \frac{d\zeta}{dt} = Q_-(t)\zeta, \tag{2.14}$$

where η=col(ξ,ζ).

Define the Cauchy matrices Φ+(t,s) and Φ−(t,s) such that Φ+(t,t)=Ek, Φ−(t,t)=−En−k, where EjE_j is a unit (j×j)-matrix.

Let us prove the following assertion about the representation of bounded solutions to system (2.1).

THEOREM 2.2 *Let all elements of* P(t) *and* g(t) *in system* (2.1) *be continuous and bounded for* t∈(−∞,+∞), *and let system* (2.2) *be hyperbolic. Then system* (2.1) *has a unique bounded solution and this solution can be represented in the form*

$$x(t) = U(t)z(t), \quad z(t) = \int_{-\infty}^{+\infty} G(t,s)h(s)\,ds, \tag{2.15}$$

where

$$G(t,s) = \begin{cases} \operatorname{diag}\{\Phi_+(t,s), 0_{n-k}\}, & t > s, \\ \operatorname{diag}\{0_k, \Phi_-(t,s)\}, & t < s, \end{cases} \tag{2.16}$$

$$G(s+0,s)-G(s-0,s)=E_n,$$

$$h(t)=U^{-1}(t)g(t)=\{h_+(t),h_-(t)\}. \qquad (2.17)$$

Proof Let us substitute

$$y(t)=U(t)z(t) \qquad (2.18)$$

into system (2.1), then we get the system

$$\frac{dz}{dt}=Q(t)z+h(t), \qquad (2.19)$$

for which the homogeneous system is of the form (2.13), (2.14).

Consider the matrix (2.16). It follows from the properties of the matrices Φ+(t), Φ−(t) that equality (2.17) is fulfilled. It follows from hyperbolicity of system (2.14) that

$$\|\Phi_+(t,s)\|\le ae^{-\lambda(t-s)}, \quad t>s,$$

$$\|\Phi_-(t,s)\|\le ae^{\lambda(t-s)}, \quad t<s. \qquad (2.20)$$

It follows from (2.16) and (2.20) that the integral in (2.15) converges for bounded functions h(t) every t. Computing the derivative of Green's matrix, we get

$$\frac{dG(t,s)}{dt} = Q(t)G(t,s). \tag{2.21}$$

Let us verify now that formula (2.15) defines the solution of equation (2.19). Representing z(t)$z(t)$ in the form

$$z(t) = \int_{-\infty}^{t} G(t,s)h(s)\,ds + \int_{t}^{+\infty} G(t,s)h(s)\,ds,$$

differentiating it and taking into account (2.21) and (2.17), we get

$$\frac{dz}{dt} = \int_{-\infty}^{t} Q(t)G(t,s)h(s)\,ds + \int_{t}^{+\infty} Q(t)G(t,s)h(s)\,ds$$
$$+ \left[G(s+0,s) - G(s-0,s)\right]h(t) = Q(t)z(t) + h(t).$$

The obtained solution is unique. If we assume the existence of two bounded solutions z1 and z2, then z1−z2 is a bounded on the axis solution of (2.2). From hyperbolicity, it follows that it is a zero solution.

COROLLARY 2.1 If homogeneous system *(2.2)* is hyperbolic and *k=n (k=0)*, then nonhomogeneous system *(2.1)* has a unique bounded for *t∈(−∞,+∞)* solution, and this solution can be represented in the following form:

$$x(t) = U(t)z(t), \quad z(t) = \int_{-\infty}^{t} G(t,s)h(s)\,ds \left(z(t) = -\int_{t}^{+\infty} G(t,s)h(s)\,ds \right), \tag{2.22}$$

where $G(t,s)=\Phi(t)\Phi^{-1}(s)$, $\Phi(t)$ *is a fundamental matrix of system* (2.2).

REMARK 2.1 If the matrix P(t)$P(t)$ in (2.1) is a constant one, analogous results are obtained in [29, 30]. The existence of a unique bounded solution under the assumption of the exponential dichotomy on (−∞,+∞)$(-\infty,+\infty)$ for system (2.2) with bounded variable coefficients is known (see, [27], p.69, Proposition 2). Similar topics were also studied in [31].

3 RESULTS: ABOUT REDUCTION OF INTEGRO-DIFFERENTIAL EQUATIONS TO SYSTEMS OF ORDINARY DIFFERENTIAL EQUATIONS

3.1 Reduction to the system of first-order ordinary differential equations

Consider the system

$$x'(t) = X\left(t, x, \int_{-\infty}^{t} K(t,s) g\big(s, x(s)\big)\,ds \right), \tag{3.1}$$

where the kernel K(t,s) is of the form

$$K(t,s) = \sum_{i,j=1}^{\infty} C_j \Phi_j(t) R_j(s) = \sum_{i,j=1}^{\infty} C_j K_j(t,s). \tag{3.2}$$

Series in (3.2) can be, for example, corresponding orthogonal expansions, series of exponents. One of the interesting cases is a finite sum in (3.2). We assume that all the matrices Φj(t)$\Phi_j(t)$ are differentiable and invertible. We can write

$$K_j(t,s) = \Phi_j(t)\Phi_j^{-1}(s)K_j(s,s). \tag{3.3}$$

Define the so-called multiplicative derivative [32]

$$P_j(t) = \frac{d\Phi_j(t)}{dt}\Phi_j^{-1}(t), \tag{3.4}$$

the matrix

$$G_j(t,s) = \Phi_j(t)\Phi_j^{-1}(s), \tag{3.5}$$

is the Cauchy matrix of the system

$$w_j'(t) = P_j(t)w(t). \tag{3.6}$$

Let us set

$$z_j(t,t_0) = \int_{-\infty}^{t} G_j(t,s)K_j(s,s)g(s,x(s))\,ds = \int_{-\infty}^{t} G_j(t,s)h_j(s,x(s))\,ds, \tag{3.7}$$

where hj(s,x(s))=Kj(s,s)g(s,x(s)).

If the matrix Gj(t,s), defined by (3.5), satisfies the inequality

$$|G_j(t,s)| \le ae^{-\lambda(t-s)}, \quad t \ge s$$

(this is the analog of (2.6)) for $k = n$, then $z_j(t, t_0)$ in formula (3.7), according to Corollary 2.1, can be considered as a solution of the one-point problem

$$z_j'(t) = P_j(t)z(t) + h_j\big(t, x(t)\big),$$

$$z_j(t_0) = z_j^0, \tag{3.8}$$

where

$$z_j^0 = \int_{-\infty}^{t_0} G_j(t_0, s)h_j\big(s, x(s)\big)\, ds,$$

if we consider $x(t)$ as a known function bounded on $(-\infty, t_0]$. Adding to equation (3.8) the so-called initial function (continuous and bounded on $(-\infty, t_0]$)

$$x(t) = \varphi(t), \tag{3.9}$$

we can consider representation (3.7) as a substitution, which leads us to the one-point problem

$$z_j'(t) = P_j(t)z(t) + h_j\big(t, x(t)\big), \quad z_j(t_0) = z_j^0, \tag{3.10}$$

where z_j^0 was defined above.

We have proven the following assertion.

THEOREM 3.1 *Let*

1. matrices $\Phi_j(t)$ in the kernels *(3.2)* be continuously differentiable and invertible for $t \in (-\infty, +\infty)$, $j=1,2,\ldots$,
2. systems *(3.6)* be of dimension n_j, where

$$P_j(t) = \frac{d\Phi_j(t)}{dt}\Phi_j^{-1}(t),$$

be hyperbolic for every j in the sense of Definition *2.1* (for $k=n$).

Then the bounded solution $x(t) \in R^n$ of system *(3.1)* with the kernel of the form *(3.2)* and the initial function *(3.9)* and the first component $x(t) \in R^n$ of the solution to the countable system

$$\begin{aligned} &x'(t) = X(t, x, z_1, z_2, \ldots), \\ &z_j'(t) = P_j(t)z(t) + h_j(t,x), \quad t \in [t_0, +\infty), \\ &z_j(t_0) = z_j^0, \qquad x(t_0) = \varphi(t_0), \end{aligned} \tag{3.11}$$

where $x \in R^n$, $z_j \in R^n$, $h_j(t,x) = K_j(t,t)g(t,x)$,

$$z_j^0 = \int_{-\infty}^{t_0} G_j(t_0, s)h_j(s, x(s))\,ds, \quad j = 1, 2, 3, \ldots$$

coincide.

REMARK 3.1 If (3.2) is a finite sum, then system (3.11) is finite dimensional.

REMARK 3.2 The system of the form

$$x'(t) = X\left(t,x,\int_0^{\infty} K(\xi)g(\xi,x(t-\xi))\,d\xi\right) \tag{3.12}$$

can be found in various applications. It can be reduced by the change of variable t−ξ=s to system (3.1) with the kernel K(t,s)=K(t−s).

REMARK 3.3 System (3.11) can be used for studying qualitative properties and for an approximate solution of system (3.1) of integro-differential system (3.1). An important basis is the theory of countable systems [33–37]; see also the papers [24, 38–41].

REMARK 3.4 Analogous result could be obtained for the system

$$x'(t) = X\left(t,x,\int_t^{+\infty} K(t,s)g(s,x(s))\,ds\right),$$

in Section 5.

3.2 Reduction to the system of ordinary differential equations of high orders

Consider the nonhomogeneous linear equation of *n* th order

$$L[y] \equiv y^{(n)} + p_1(t)y^{(n-1)} + \cdots + p_n(t)y = f(t) \tag{3.13}$$

and the corresponding homogeneous equation

$$L[z] \equiv z^{(n)} + p_1(t)z^{(n-1)} + \cdots + p_n(t)z = 0, \tag{3.14}$$

where all coefficients pj (j=1,...,n) and f are essentially bounded on (−∞,+∞).

Let

$$z_1(t), \ldots, z_n(t) \tag{3.15}$$

be a fundamental system of solutions of equation (3.14). Using (3.15), we can construct the solution z=ψ(t,t0) such that

$$z(t_0) = \psi(t_0, t_0) = 0, \qquad \ldots, \qquad z^{(n-2)}(t_0) = 0, \qquad z^{(n-1)}(t_0) = 1. \tag{3.16}$$

The function ψ(t,s) is called the Cauchy function of equation (3.13) [42, 43]. Consider the function

$$y(t) = \int_{-\infty}^{t} \psi(t,s)f(s)\,ds, \tag{3.17}$$

assuming that the integral converges. Let us verify that (3.17) is a solution of (3.13). Actually,

$$y^{(k)}(t) = \int_{-\infty}^{t} \psi^{(k)}(t,s)f(s)\,ds + \psi^{(k-1)}(t,t)f(t) = \int_{-\infty}^{t} \psi^{(k)}(t,s)f(s)\,ds \tag{3.18}$$

for k=1,...,n−1$k = 1, \ldots, n-1$, and

$$y^{(n)}(t) = \int_{-\infty}^{t} \psi^{(k)}(t,s)f(s)\,ds + f(t). \tag{3.19}$$

It follows from (3.18), (3.19) and the equality L[ψ(t,t0)]=0$L[\psi(t,t_0)] = 0$ that

$$L[y] = f(t). \tag{3.20}$$

The obtained particular solution satisfies the initial conditions

$$y(t_0) = \int_{-\infty}^{t_0} \psi(t_0, s) f(s)\, ds,$$
$$y^{(k)}(t_0) = \int_{-\infty}^{t_0} \psi^{(k)}(t_0, s) f(s)\, ds, \quad k = 1, \ldots, n-1. \tag{3.21}$$

EXAMPLE 3.1 For the equation

$$y^{(n)}(t) = f(t), \tag{3.22}$$

we get

$$\psi(t,s) = \frac{1}{(n-1)!}(t-s)^{n-1},$$
$$y(t) = \frac{1}{(n-1)!} \int_{-\infty}^{t} (t-s)^{n-1} f(s)\, ds. \tag{3.23}$$

EXAMPLE 3.2 For the equation

$$\sum_{j=0}^{n} c_n^j \lambda^j y^{(n-j)}(t) = f(t), \tag{3.24}$$

we get

$$\psi(t,s) = \frac{1}{(n-1)!}(t-s)^{n-1}e^{-\lambda(t-s)},$$

$$y(t) = \frac{1}{(n-1)!}\int_{-\infty}^{t}(t-s)^{n-1}e^{-\lambda(t-s)}f(s)\,ds. \quad (3.25)$$

Consider the system

$$x'(t) = X\left(t, x(t), \int_{-\infty}^{t} F(t,s,y(s))\,ds\right), \quad t \in [t_0, +\infty), \quad (3.26)$$

where x=col(u,y) $x = \mathrm{col}(u,y)$, u∈Rn-1, y∈R1, and assume that

$$F(t,s,y(s)) = \sum_{j=1}^{\infty} \psi_j(t,s) g_j(s,y(s)), \quad (3.27)$$

where ψj(t,s) (j=1,2,...) are the Cauchy functions of corresponding linear equations

$$L_j[z_j] \equiv z_j^{(n)} + p_{1j}(t)z_j^{(n-1)} + \cdots + p_{nj}(t)z_j = 0, \quad j = 1,2,\ldots. \quad (3.28)$$

Define

$$v_j(t) = \int_{-\infty}^{t} \psi_j(t,s) g_j(s,y(s))\,ds, \quad (3.29)$$

and set

$$y(t) = \varphi(t), \quad t \in (-\infty, t_0], \tag{3.30}$$

where φ is considered as a known function. Assuming that the Cauchy functions imply convergence of integrals (3.29) for all j and that the function φ is bounded, we obtained that system (3.26) is reduced to the countable system

$$\begin{aligned}
&x'(t) = X(t, x, v_1, v_2, \ldots),\\
&L_j[v_j] = g_j(t, y),\\
&x(t_0) = x_0,\\
&v_j^{(k)}(t_0) = \int_{-\infty}^{t} \psi_j^{(k)}(t, s) g_j(s, \varphi(s))\, ds,\\
&\quad t \in [t_0, +\infty), k = 0, \ldots, n_j, j = 1, 2, \ldots
\end{aligned} \tag{3.31}$$

in the sense that the solution of (3.26) coincides with the component x of the solution vector of (3.31).

4 RESULTS: EXAMPLES OF REDUCTION OF INTEGRO-DIFFERENTIAL EQUATIONS TO SYSTEMS OF ORDINARY DIFFERENTIAL EQUATIONS

EXAMPLE 4.1 Model of tumor-immune system [9]

$$x'(t) = x(f(x) - \Phi(x, y)),$$

$$y' = \beta(z)y - \mu(x)y + \sigma q(x) + \theta(t),$$

$$z = \int_{-\infty}^{t} K(t - s)x(s)\, ds. \qquad (4.1)$$

This system is an example of two-dimensional system (3.26) with distributed delay of x. In [9] the following kernel

$$K(t) = Erl_{\lambda,n}t = \frac{\lambda^n}{(n-1)!}t^{n-1}e^{-\lambda t}, \qquad (4.2)$$

is used and the case n=1$n = 1$ is studied in detail. It is clear from Example 3.1 (see (3.24) and (3.25)) that the substitution

$$z(t) = \frac{1}{(n-1)!}\int_{-\infty}^{t}(t-s)^{n-1}e^{-\lambda(t-s)}x(s)\, ds \qquad (4.3)$$

reduces system (4.1) with the kernel (4.2) to the system of ordinary differential equations

$$x'(t) = x(f(x) - \Phi(x, y)),$$

$$y' = \beta(z)y - \mu(x)y + \sigma q(x) + \theta(t),$$

$$\sum_{j=0}^{n} c_n^j \lambda^j z^{(n-j)}(t) = \lambda^n x(t), \qquad (4.4)$$

which can be written as a system of the order n+2. Note that for n=1$n = 1$, the last equation in system (4.4) is of the form z'+z=λx.

EXAMPLE 4.2 Model of hematopoiesis [10]. This model can be written in the form

$$P'(t) = -\delta(t)P(t) - \frac{\beta(t)P(t)}{1+P^n(t)} + \alpha(t)\int_0^\infty K(\tau)\frac{P(t-\tau)}{1+P^n(t-\tau)}\,d\tau. \tag{4.5}$$

The coefficients α, β and δ in (4.5) are positive ω-periodic functions, the kernel K satisfies the condition $\int_0^\infty K(\tau)\,d\tau = 1$. The change of variable t−τ=s and then the substitution of the type (4.3) in the case of the kernel (4.2) reduces integro-differential equation (4.5) to the system of ordinary differential equations

$$P'(t) = -\delta(t)P(t) - \frac{\beta(t)P(t)}{1+P^n(t)} + \alpha(t)z(t),$$

$$\sum_{j=0}^{n} c_n^j \lambda^j z^{(n-j)}(t) = \lambda^n \frac{P(t)}{1+P^n(t)}. \tag{4.6}$$

Consider now the case of both distributed and concentrated delays in the system

$$y'(t) = Y\left(t, x(t), \int_{-\infty}^{t} K(t,s)q(s,y(s))\,ds\right),$$

$$z'(t) = Z(t, x(t), z(t-\tau)). \tag{4.7}$$

Let us describe the process of reduction, which is similar to the process described in the Section 3.1. The vector x is of the form x=col(y,z). Denoting

$$v(t) = \int_{-\infty}^{t} K(t,s)q(s,y(s))\,ds, \tag{4.8}$$

we make the substitution (compare with (3.7))

$$v(t) = \int_{-\infty}^{t} G_j(t,s)h_j(s,y(s))\,ds. \tag{4.9}$$

Introduce the initial functions

$$y(t) = \varphi(t), \quad t \in (-\infty, t_0], \qquad z(t) = \psi(t), \quad t \in (-\tau, 0]. \tag{4.10}$$

Under the assumption of convergence of the integrals, substitution (4.8) reduces (4.7) to the system

$$\begin{aligned} y'(t) &= Y\big(t, x(t), v_1(t), v_2(t), \ldots\big), \\ z'(t) &= Z\big(t, x(t), z(t-\tau)\big), \\ v_j'(t) &= P_j(t)v_j(t) + h_j\big(t, y(t)\big), \quad j = 1, 2, \ldots \end{aligned} \tag{4.11}$$

with the initial conditions defined by (4.10).

EXAMPLE 4.3 The model of the plankton-nutrient interaction [25]

$$\begin{aligned} N'(t) &= D\big(N^0 - N(t)\big) - aP(t)U\big(N(t)\big) + \gamma \int_{-\infty}^{t} F(t-s)P(s)\,ds, \\ P'(t) &= P\big\{a_1 U\big(N(t-\tau)\big) - (y + D_1)\big\}. \end{aligned} \tag{4.12}$$

The initial functions

$$P(t) = \psi(t), \quad t \in (-\infty, t_0],$$

$$N(t) = \varphi(t), \quad t \in (-\tau, 0]. \tag{4.13}$$

The description of all parameters can be found in the paper [25]. U(N)$U(N)$ is a known function. Concerning the function U(N)$U(N)$, it is assumed that

$$U(0) = 0, \quad \frac{dU}{dN} > 0, \quad \lim_{N \to \infty} U(N) = 1. \tag{4.14}$$

A particular case of U(N) is

$$U(N) = \frac{N}{k + N}, \quad \text{where } k > 0. \tag{4.15}$$

Concerning the kernel, it is assumed that F(t) is a bounded nonnegative function such that $\int_{-\infty}^{+\infty} F(t)\,dt = 1.$

In [25] the properties of system (4.12) are considered in various particular cases of the kernel F(t)$F(t)$. The most general of them is the following:

$$F(t) = \alpha e^{-\alpha t}, \quad \text{where } \alpha > 0. \tag{4.16}$$

It is clear from (4.2) for n=1$n = 1$ that

$$F(t) = Erl_{\alpha,1}(t). \tag{4.17}$$

The substitution (4.3) for n=1$n = 1$ is of the form

$$z(t) = \int_{-\infty}^{t} e^{-\alpha(t-s)} P(s)\, ds \tag{4.18}$$

and it reduces system (4.12) to the system

$$\begin{aligned} N'(t) &= D\left(N^0 - N(t)\right) - aP(t)U\left(N(t)\right) + \gamma z(t), \\ P'(t) &= P\left\{a_1 U\left(N(t-\tau)\right) - (y + D_1)\right\}, \\ z'(t) + z(t) &= \alpha P(t). \end{aligned} \tag{4.19}$$

5 RESULTS: SYSTEMS WITH ADVANCED ARGUMENT

Using results of Section 2 and approach of Section 3 (Section 3.1), we describe reduction of the integro-differential system

$$x'(t) = X\left(t, x, \int_{t}^{+\infty} K(t,s) g\left(s, x(s)\right) ds\right), \tag{5.1}$$

and the kernel

$$K(t,s) = \sum_{i,j=1}^{\infty} C_j \Phi_j(t) R_j(s) = \sum_{i,j=1}^{\infty} C_j K_j(t,s) \tag{5.2}$$

to a system of ordinary differential equations. Introducing the matrix Gj(t,s) and the equation

$$w_j'(t) = P_j(t)w(t),$$

by the formulas (3.3), (3.4) and (3.5), let us require that Gj(t,s)$G_j(t,s)$ (j=1,2,...$j=1,2,\dots$) satisfy inequalities (2.6) under the assumption that k=0$k=0$ in the condition of hyperbolicity.

Introduce the substitution

$$z_j(t,t_0) = -\int_t^{+\infty} G_j(t,s)h_j\big(s,x(s)\big)\,ds. \tag{5.3}$$

According to Corollary 2.1, we can consider (5.3) for k=0$k=0$ as the solution of the one-point problem for system (3.8), supposing x(t)$x(t)$ is a known function

$$x(t) = \psi(t), \quad t \in [t_0, +\infty). \tag{5.4}$$

As a result, we obtain an analog of Theorem 3.1 for equation (5.1) with the kernel (5.2).

THEOREM 5.1 *Let*

1. (a) matrices Φj(t) in the kernels (5.2) be continuously differentiable and invertible for t∈(−∞,+∞), j=1,2,...$j=1,2,\dots$,

(b) systems (3.6) be of dimension njn_j, where

$$P_j(t) = \frac{d\Phi_j(t)}{dt}\Phi_j^{-1}(t),$$

be hyperbolic for every j in the sense of Definition 2.1 (for k=n$k = n$).

Then the bounded solution x(t)∈R^n of system (5.1) with the kernel of the form (5.2) and the end function (5.4) and the first component x(t)∈R^n of the solution to the system

$$x'(t) = X(t, x, z_1, z_2, \ldots),$$

$$z_j'(t) = P_j(t)z(t) + h_j(t,x), \quad t \in [t_0, +\infty),$$

$$z_j(t_0) = z_j^0, \qquad x(t_0) = \psi(t_0), \tag{5.5}$$

where x∈R^n, zj∈Rn, hj(t,x)=Kj(t,t)g(t,x),

$$z_j^0 = -\int_{-\infty}^{t_0} G_j(t_0,s)h_j(s,x(s))\,ds, \quad j = 1,2,3,\ldots$$

coincide.

6 RESULTS: ABOUT SYSTEMS WITH BOTH DELAYED AND ADVANCED ARGUMENT

Let us consider system (1.2) with distributed delay and advance. Denote x=col(y,z)$x = \mathrm{col}(y,z)$ in such a form that system (1.2) can be written in the form

$$y'(t) = Y\left(t, x(t), \int_{-\infty}^{t} K_1(t,s)g_1(s,y(s))\,ds\right),$$

$$z'(t) = Z\left(t, x(t), \int_{t}^{+\infty} K_2(t,s)g_2(s,z(s))\,ds\right). \tag{6.1}$$

Using the technique of Sections 3 and 5, we introduce

$$u_j(t) = \int_{-\infty}^{t} G_j^1(t,s)h_j^1\left(s,x(s)\right) ds, \tag{6.2}$$

$$v_j(t) = -\int_{t}^{+\infty} G_j^2(t,s)h_j^2\left(s,x(s)\right) ds, \tag{6.3}$$

where $h_j^i(t,x) = K_i(t,t)g_j(t,x),\ i = 1,2,\ j = 1,2,\ldots$ and denote $P_j^1(t) = \frac{d\Phi_j}{dt}\Phi_j^{-1}(t),\ j = 1,2,\ldots,\ P_r^2(t) = \frac{d\Phi_r}{dt}\Phi_r^{-1}(t),\ r = 1,2,\ldots.$

Requiring $G_j^i(t,s)$ satisfies inequality (2.6), the first for k=n and the second for k=0, we obtain that solution x=col(y,z) of system (6.1) satisfies also the following problem:

$$y'(t) = Y\left(t,x(t),u_1(t),u_2(t),\ldots\right),$$

$$z'(t) = Z\left(t,x(t),v_1(t),v_2(t),\ldots\right),$$

$$u_j'(t) = P_j^1(t)u_j(t) + h_j^1\left(t,y(t)\right), \quad j = 1,2,\ldots,$$

$$v_r'(t) = P_r^2(t)u_r(t) + h_r^1\left(t,z(t)\right), \quad r = 1,2,\ldots \tag{6.4}$$

with conditions

$$y(t_0) = \varphi(t_0), \qquad u_j(t_0) = \int_{-\infty}^{t_0} G_j^1(t_0,s)h_j^1\left(s,x(s)\right) ds, \quad j = 1,2,\ldots,$$

$$z'(t) = \psi(t_0), \qquad v_r(t_0) = \int_{t_0}^{+\infty} G_r^2(t_0,s)h_r^2\left(s,x(s)\right) ds, \quad r = 1,2,\ldots,$$

$$y(t) = \varphi(t), \quad t \in (-\infty,t_0),$$

$$z(t) = \psi(t), \quad t \in (t_0 < +\infty). \tag{6.5}$$

7 CONCLUSIONS

The method described above allows us to reduce systems of integro-differential systems with distributed delay and/or advance to systems of ordinary differential equations. For Volterra systems of the type (1.1), it was a basis for studying stability, bifurcation, Floquet theory, parametric resonance, stabilization and oscillation properties for integro-differential equations with ordinary [13–18, 20] and partial [21, 22] derivatives. We could extend the main results of these works to integro-differential equation (1.2).

Generally speaking, after the reduction, we get infinity dimensional systems of ordinary differential equations. For their analysis, the theory of countable differentiable systems could be used [33–37].

In the study of various biological systems, the linear chain trick method was used (see, for example, [9, 10]). It is clear (see Section 4) that our approach includes the linear trick method. Note also the use of W-transform, which also allows researchers to reduce integro-differential equations to systems of ordinary differential equations [11].

The proposed method allows us also to study generalized and impulsive systems. For example, in the case of discontinuous solutions described by Heaviside functions $H_a(t)$ we can use its connection with δ-function: $H_a(t) = \int_{-\infty}^{t} \delta_a(\xi)\, d\xi$ and to get to a system of integro-differential equations. Introducing the sequence, for example,

$$Erl_{\lambda,n}t = \frac{\lambda^n}{(n-1)!} t^{n-1} e^{-\lambda t},$$

where $\lambda \to \infty$, we can consider the obtained system of integro-differential equations as an approximation of generalized equations.

This allows us in corresponding cases to reduce the study of a generalized and impulsive system to the analysis of the sequence of integro-differential equations, and consequently to the analysis of the corresponding sequence of systems of ordinary differential equations.

ACKNOWLEDGEMENTS

Authors thank the reviewers for their reports, which have essentially improved the paper.

COMPETING INTERESTS

There are no competing interests.

AUTHORS' CONTRIBUTIONS

Results are obtained by both authors as a result of their many years of collaboration. All authors read and approved the final manuscript.

REFERENCES

1. Drozdov AD, Kolmanovskii VB: Stability in Viscoelasticity. North Holland, Amsterdam; 1994.
2. Fabrizio M, Morro A SIAM Stud. Appl. Math. In Mathematical Problems in Linear Viscoelasticity. SIAM, Philadelphia; 1992.

3. Golden JM, Graham GAC: Boundary Value Problems in Linear Viscoelasticity. Springer, Berlin; 1988.

4. Gurtin ME, Pipkin AC: A general theory of heat conduction with finite wave speeds. Arch. Ration. Mech. Anal. 1968, 31: 113-126.

5. Novick-Cohen A: Conserved phase-field equations with memory. GAKUTO Internat. Ser. Math. Sci. Appl 5. In Curvature Flows and Related Topics. Edited by: Damlamian A, Spruck J, Visintin A. Gakkotosho, Tokyo; 1995:179-197.

6. Corduneanu C: Integral Equations and Stability Feedback Systems. Academic Press, New York; 1973.

7. Corduneanu C: Integral Equations and Applications. Cambridge University Press, Cambridge; 1991.

8. Burton TA: Volterra Integral and Differential Equations. Academic Press, New York; 1983.

9. D'Onofrio A, Gatti F, Cerrai P, Freshi L: Delay-induced oscillatory dynamics of tumor-immune system interaction. Math. Comput. Model.2010, 52: 572-591.

10. Weng P-X: Global attractive periodic solution in a model of hematopoiesis. Comput. Math. Appl. 2002, 44: 1019-1030.

11. Ponosov A, Shindiapin A, Miguel J: The W-transform links delay and ordinary differential equations. Funct. Differ. Equ. 2002, 9: 437-470.

12. Domoshnitsky A: Exponential stability of convolution integro-differential equations. Funct. Differ. Equ. 1998, 5: 445-455.

13. Domoshnitsky A, Gotser Y: Hopf bifurcation of integro-differential equations. Electron. J. Qual. Theory Differ. Equ. 2000, 3: 1-11.

14. Domoshnitsky A, Goltser Y: Approach to study of bifurcations and stability of integro-differential equations. Math. Comput. Model. 2002, 36: 663-678.

15. Domoshnitsky A, Goltser Y: On stability and boundary value problems for integro-differential equations. Nonlinear Anal., Theory Methods Appl. 2005, 63(5-7):e761-e767.

16. Goltser Y, Domoshnitsky A: Bifurcation and stability of integro-differential equations. Nonlinear Anal., Theory Methods Appl. 2001, 47: 953-967. 1

17. Agarwal RP, Bohner M, Domoshnitsky A, Goltser Ya: Floquet theory and stability of nonlinear integro-differential equations. Acta Math. Hung. 2005, 109(4):305-330.

18. Goltser Y, Agronovich G: Normal form and parametric resonance in linear almost periodic systems. Funct. Differ. Equ. 2002, 9: 117-134.

19. Goltser Y: On bifurcation of steady resonance modes for nonlinear integro-differential equations. Differ. Equ. Dyn. Syst. 2006, 2: 239-251

20. Domoshnitsky A, Maghakyan A, Puzanov N: About stabilization by feedback control in integral form. Georgian Math. J. 2012. 10.1515/gmj-2012-0033

21. Agarwal RP, Domoshnitsky A, Goltser Y: Stability of partial functional integro-differential equations. J. Dyn. Control Syst. 2006, 12: 1-31.

22. Domoshnitsky A, Goltser Y: Constructive approach to phase transition model. Funct. Differ. Equ. 2000, 7: 269-278.

23. Goltser Y, Litsyn L: Volterra integro-differential systems of ordinary differential equations. Math. Comput. Model. 2005, 42: 221-233.

24. Goltser Y, Litsyn L: Non-linear Volterra IDE, infinite systems and normal form of ODE. Nonlinear Anal., Theory Methods Appl. 2008, 68: 1553-1569.

25. Shingui R: The effect of delays on stability and persistence in plankton models. Nonlinear Anal., Theory Methods Appl. 1995, 24: 575-585.

26. Filatov AN: Averaging in Differential and Integro-Differential Equations. FAN, Tashkent; 1971. (in Russian)

27. Coppel WA Lecture Notes in Mathematics 629. In Dichotomies in Stability Theory. Springer, Berlin; 1978.

28. Pliss VA: Integral Sets of Periodic Systems of Differential Equations. Nauka, Moscow; 1977.

29. Bibikov YN: Course of Ordinary Differential Equations. Vysshaya Shkola, Moscow; 1991.

30. Demidovich BP: Lectures on the Mathematical Theory of Stability. Nauka, Moscow; 1967.

31. Mitropolsky YA, Samoilenko AM, Kulik VL Stability and Control: Theory, Methods and Applications 14. In Dichotomies and Stability in Nonautonomous Linear Systems. Taylor & Francis, London; 2003.

32. Gantmacher FR: Theory of Matrices. Chelsea Publishing, New York; 1959.

33. Persidskii KP 2. In Selected Works. Nauka, Alma-Ata; 1976.

34. Persidskii KP: On stability of solutions to a countable systems of differential equations. Izv. AS Kaz. SSR, Math. Mech. 1948, 2: 2-35.

35. Persidskii KP: Infinite systems of differential equations. Izv. AS Kaz. SSR, Math. Mech. 1956, 4(8):3-11.

36. Persidskii KP: Countable systems of differential equations and stability of their solutions. Izv. AS Kaz. SSR, Math. Mech. 1959, 7(11):52-71.

37. Valeev KG, Zhautykov OA: Infinite Systems of Differential Equations. Nauka, Alma-Ata; 1974.

38. Domoshnitsky A, Litsyn E: Positivity of the Green's matrix of infinite system. Panam. Math. J. 2006, 16(2):27-40.

39. Domoshnitsky A, Goltser Y: Positivity of solutions to boundary value problems for infinite functional differential systems. Math. Comput. Model. 2007, 45(11-12):1395-1404.

40. Litsyn L: On the formula for general solution of infinite system of functional differential equations. Funct. Differ. Equ. 1994, 2: 111-121.

41. Litsyn L: On the general theory of liner functional differential equations. Differ. Uravn. (Minsk) 1988, 24: 977-986.

42. Goursat E: Cours d'analyse mathematique. Gauthier-Villars, Paris; 1925.

43. Stepanov, VV: Course of Differential Equations. Moscow (1958)

Index